WASTEWATER AND PUBLIC HEALTH

Bacterial and Pharmaceutical Exposures

WASTEWATER AND PUBLIC HEALTH

Bacterial and Pharmaceutical Exposures

Edited by
Victor M. Monsalvo

Apple Academic Press Inc. | Apple Academic Press Inc.
3333 Mistwell Crescent | 9 Spinnaker Way
Oakville, ON L6L 0A2 | Waretown, NJ 08758
Canada | USA

©2016 by Apple Academic Press, Inc.

First issued in paperback 2021

Exclusive worldwide distribution by CRC Press, a member of Taylor & Francis Group

No claim to original U.S. Government works

ISBN 13: 978-1-77463-531-5 (pbk)
ISBN 13: 978-1-77188-164-7 (hbk)

Library and Archives Canada Cataloguing in Publication

Wastewater and public health : bacterial and pharmaceutical exposures / edited by Victor M. Monsalvo.

Includes bibliographical references and index.
ISBN 978-1-77188-164-7 (bound)
1. Sewage--Purification--Health aspects. 2. Bacterial pollution of water. 3. Drugs--Environmental aspects. 4. Public health. I. Monsalvo, Victor (Victor M.), editor

RA591.W28 2015   363.72'84   C2015-902239-8

Library of Congress Cataloging-in-Publication Data

Wastewater and public health : bacterial and pharmaceutical exposures / Victor M. Monsalvo, editor.

pages cm
Includes bibliographical references and index.
ISBN 978-1-77188-164-7 (alk. paper)
1. Drugs--Environmental aspects. 2. Bacterial pollution of water. 3. Water--Microbiology. 4. Water--Pollution. I. Monsalvo, Victor M., editor.

TD196.D78W37 2015   628.4'3--dc23   2015011037

Apple Academic Press also publishes its books in a variety of electronic formats. Some content that appears in print may not be available in electronic format. For information about Apple Academic Press products, visit our website at **www.appleacademicpress.com** and the CRC Press website at **www.crcpress.com**

ABOUT THE EDITOR

VICTOR M. MONSALVO

Professor Victor Monsalvo is an environmental scientist with a PhD in chemical engineering from the University Autonoma de Madrid, where he later became a professor in the chemical engineering department. As a researcher, he has worked with the following universities: Leeds, Cranfield, Sydney, and Aachen. He is part of an active research team working in areas of environmental technologies, water recycling, and advanced biological systems, including membrane bioreactors. He has led eight research projects with private companies and an R&D national project, coauthored two patents (national and international), and written around fifty journal and referred conference papers. He has given two key notes in international conferences and has been a member of the organizing committee of five national and international conferences, workshops, and summer schools, and is currently involved in fifteen research projects sponsored by various entities. He is currently working as senior researcher in the Chemical Processes Department at Abengoa Research.

CONTENTS

ACKNOWLEDGMENT AND
HOW TO CITE

The editor and publisher thank each of the authors who contributed to this book. The chapters in this book were previously published elsewhere. To cite the work contained in this book and to view the individual permissions, please refer to the citation at the beginning of each chapter. Each chapter was read individually and carefully selected by the editor; the result is a book that provides a multiperspective look at the ways in which wastewater management impacts public health. The chapters included examine the following topics:

- In chapter 1, the authors review the major genera of pathogenic human enteric viruses, their pathogenicities and epidemiology, as well as the role of wastewater effluents in their transmission. From this foundation, they discuss future directions for overcoming existing challenges.
- The study in chapter 2 assesses the prevalence of non-cholerae *Vibrios* in the final effluents of 14 wastewater treatment facilities in the ECP of South Africa, characterizing them into species and determining their antibiogram properties and evaluating the public health implications of the findings.
- Chapter 3 examines the prevalence of *Listeria spp.* in various wastewater and sludge samples in northern municipal wastewater treatment plant in Isfahan, Iran, and evaluates the performance of this treatment plant on removal of *Listeria spp.*
- The objective in chapter 4 was to investigate antibiotic-resistance profiles in *E. coli* isolated from two local wastewater treatment plants, with both raw and treated wastewater samples included. Data obtained in this study indicated that wastewater treatment processes, together with effective dilution of treated wastewater by marine outfall, were generally sufficient to protect coastal water quality from sanitary degradation—but human-associated bacteria, even potential pathogens and bacteria carrying antibiotic resistance genes of clinical significance, survived in wastewater and marine water conditions.
- In chapter 5, the authors used culture-independent approaches to determine the prevalence of antibiotic resistance genes (ARGs) and to examine how bacterial communities from biofilms and sediments respond to the discharge of wastewater treatment plant effluents in the receiving river. ARGs and bacterial community composition in the upstream river were

also analyzed to determine the contribution of wastewater discharge to antibiotic resistance in the downstream river samples.

- The dissemination of antibiotic-resistant bacteria and antibiotic resistant genes from wastewater irrigation to natural soil and water environments may contribute to global antibiotic resistance. In chapter 6, the authors discuss the implications of wastewater reuse in agriculture and the impact of treated wastewater irrigation.

- The authors of chapter 7 offer a comprehensive investigation of the microbial community structure and function of anaerobic and aerobic sludge in a full-scale tannery wastewater treatment plant.

- In chapter 8, the authors designed a batch experiment to culture sewage treatment plant sludge in filtered sewage fed with different concentrations of tetracycline to identify tetracycline resistant bacteria in the sludge and to evaluate the effect of tetracycline stress on the abundance and diversity of resistant genes.

- The study described in chapter 9 was aimed at screening a wastewater treatment plant for the occurrence of a large set of known antibiotic-resistance genes, using a PCR approach that would also allow for detection of low-abundance resistance genes. The study provides evidence that bacteria residing in different compartments of the treatment plant harbored various plasmid-borne resistance determinants that represented all common classes.

- Chapter 10 investigates the occurrence of representative antidepressant drugs in wastewater from wastewater treatment plants in Beijing, China, in order to facilitate a better understanding of the potential ecological and human health risks of these antidepressants and their metabolites in wastewater.

- The authors of chapter 11 investigated the removal and transformation of pharmaceuticals with regard to physicochemical and structural properties of pharmaceuticals and their metabolites in various environments, including wastewater treatment plants and constructed wetlands receiving wastewater effluent.

- In chapter 12, the authors summarize their monitoring of three groups of pharmaceutical compounds in wastewater samples The results show that the elimination of most of the analyzed compounds was incomplete, but the membrane bioreactor technique was the more efficient of the two wastewater treatment process analyzed, resulting in lower effluent concentration for most of the compounds in comparison with the activated sludge technique. The results obtained by these authors' monitoring should encourage including pharmaceutical compounds in the monitoring of wastewater effluent quality.

LIST OF CONTRIBUTORS

Martins Adefisoye
Applied and Environmental Microbiology Research Group (AEMREG), Department of Biochemistry and Microbiology, University of Fort Hare, P Bag X1314, Alice, Eastern Cape, 5700, South Africa

Cristina Afonso-Olivares
Departamento de Química, Universidad de Las Palmas de Gran Canaria, Las Palmas de Gran Canaria, 35017, Spain

Hajar Aghili
Environment Research Center, Isfahan University of Medical Sciences (IUMS), Isfahan, Iran, and Department of Environmental Health Engineering, School of Public Health, IUMS, Isfahan, Iran

Hossein Movahedian Attar
Environment Research Center, Isfahan University of Medical Sciences (IUMS), Isfahan, Iran, and Department of Environmental Health Engineering, School of Public Health, IUMS, Isfahan, Iran

Jose Luis Balcazar
Catalan Institute for Water Research (ICRA), Scientific and Technological Park of the University of Girona, Girona, Spain

Artur Burzyński
Genetics and Marine Biotechnology Department, Institute of Oceanology of the Polish Academy of Sciences, Powstancow Warszawy 55, 81-712 Sopot, Poland

Hong-Rui Chen
Research Center for Eco-Environmental Sciences, Chinese Academy of Sciences, Beijing 100085, China

J. Cho
School of Civil and Environmental Engineering, Yonsei University, Yonsei-ro 50, Seodaemun-gu, Seoul 120-749, Korea

Eddie Cytryn
Institute of Soil, Water and Environmental Sciences, Volcani Center, Agricultural Research Organization, Bet Dagan, Israel; P.O. Box 6, Bet Dagan, 50250, Israel

Wolfgang Eichler
Landesamt für Natur, Umwelt und Verbraucherschutz NRW, FB76.2, Auf dem Draap 25, 40221 Düsseldorf, Germany

Karl-Heinz Gartemann
Lehrstuhl für Gentechnologie und Mikrobiologie, Fakultät für Biologie, Universität Bielefeld, Universitätsstraße 25, D-33615 Bielefeld, Germany

Joao Gatica

Institute of Soil, Water and Environmental Sciences, Volcani Center, Agricultural Research Organization, Bet Dagan, Israel; Department of Agroecology and Plant Health, The Robert H. Smith Faculty of Agriculture, Food and Environment, The Hebrew University of Jerusalem, Jerusalem, Israel

Rayco Guedes-Alonso

Departamento de Química, Universidad de Las Palmas de Gran Canaria, Las Palmas de Gran Canaria, 35017, Spain

Siyabulela S. Gusha

Applied and Environmental Microbiology Research Group (AEMREG), Department of Biochemistry and Microbiology, University of Fort Hare, P/Bag X1314, Alice 5700, South Africa

Tim Gützkow

Institute for Genome Research and Systems Biology, Center for Biotechnology, Universität Bielefeld, Postfach 100131, D-33501 Bielefeld, Germany

Kailong Huang

State Key Laboratory of Pollution Control and Resource Reuse, School of the Environment, Nanjing University, Nanjing, China

Ying-Bin Huo

Research Center for Eco-Environmental Sciences, Chinese Academy of Sciences, Beijing 100085, China

Mohammad Jalali

Infection Disease and Tropical Medicine Research Center and School of Nutrition and Food Sciences, Isfahan University of Medical Sciences, Isfahan, Iran

Juan Jofre

Department of Microbiology, University of Barcelona, Barcelona, Spain

Y. Kim

Department of Environmental Science and Engineering, Gwangju Institute of Science and Technology (GIST), 1 Oryong-dong, Buk-gu, Gwangju, 500-712, Korea

Ewa Kotlarska

Genetics and Marine Biotechnology Department, Institute of Oceanology of the Polish Academy of Sciences, Powstancow Warszawy 55, 81-712 Sopot, Poland

Irene Krahn

Institute for Genome Research and Systems Biology, Center for Biotechnology, Universität Bielefeld, Postfach 100131, D-33501 Bielefeld, Germany

E. Lee

Department of Civil, Environmental and Architectural Engineering, University of Colorado Boulder, Boulder, CO 80309, USA

S. Lee

Woongjin Chemical Co., Ltd, KANC 906-10, lui-dong, Yeongtong-gu, Suwon-si, Gyeonggi-do 443-270, Korea

Aimin Liv

State Key Laboratory of Pollution Control and Resource Reuse, School of the Environment, Nanjing University, Nanjing, China

Burkhard Linke
Bioinformatics Resource Facility, Center for Biotechnology, Universität Bielefeld, Postfach 100131, D-33501 Bielefeld, Germany

Bo Liu
State Key Laboratory of Pollution Control and Resource Reuse, School of the Environment, Nanjing University, Nanjing, China

Chao Long
State Key Laboratory of Pollution Control and Resource Reuse, School of the Environment, Nanjing University, Nanjing, China

Aneta Łuczkiewicz
Department of Water and Wastewater Technology, Faculty of Civil and Environmental Engineering, Gdansk University of Technology, Narutowicza 11/12, 80-233 Gdansk, Poland

Elisabet Marti
Catalan Institute for Water Research (ICRA), Scientific and Technological Park of the University of Girona, Girona, Spain

Yu Miao
State Key Laboratory of Pollution Control and Resource Reuse, School of the Environment, Nanjing University, Nanjing, China

Sarah Montesdeoca-Esponda
Departamento de Química, Universidad de Las Palmas de Gran Canaria, Las Palmas de Gran Canaria, 35017, Spain

Nahid Navidjouy
Department of Environmental Health Engineering, Urmia University of Medical Sciences, Urmia, Iran

Nolonwabo Nontongana
Applied and Environmental Microbiology Research Group (AEMREG), Department of Biochemistry and Microbiology, University of Fort Hare, P Bag X1314, Alice, Eastern Cape, 5700, South Africa

Vuyokazi Nongogo
Applied and Environmental Microbiology Research Group (AEMREG), Department of Biochemistry and Microbiology, University of Fort Hare, P Bag X1314, Alice, Eastern Cape, 5700, South Africa

Anthony I. Okoh
Applied and Environmental Microbiology Research Group (AEMREG), Department of Biochemistry and Microbiology, University of Fort Hare, P Bag X1314, Alice, Eastern Cape, 5700, South Africa

Osuolale O. Olayemi
Applied and Environmental Microbiology Research Group (AEMREG), Department of Biochemistry and Microbiology, University of Fort Hare, P Bag X1314, Alice, Eastern Cape, 5700, South Africa

J. Park
School of Civil and Environmental Engineering, Yonsei University, Yonsei-ro 50, Seodaemun-gu, Seoul 120-749, Korea

Marta Pisowacka
Instituto Gulbenkian de Ciência, Rua da Quinta Grande 6, 2780-156 Oeiras, Portugal

Alfred Pühler

Institute for Genome Research and Systems Biology, Center for Biotechnology, Universität Bielefeld, Postfach 100131, D-33501 Bielefeld, Germany

Hongqiang Ren

State Key Laboratory of Pollution Control and Resource Reuse, Environmental Health Research Center, School of the Environment, Nanjing University, Nanjing 210023, China

José Juan Santana-Rodríguez

Departamento de Química, Universidad de Las Palmas de Gran Canaria, Las Palmas de Gran Canaria, 35017, Spain

Andreas Schlüter

Institute for Genome Research and Systems Biology, Center for Biotechnology, Universität Bielefeld, Postfach 100131, D-33501 Bielefeld, Germany

Ling-Hui Sheng

College of Resources and Environment, Graduate University of the Chinese Academy of Sciences, Beijing 100049, China; National Institute of Metrology, Beijing 100013, China

Peng Shi

State Key Laboratory of Pollution Control and Resource Reuse, School of the Environment, Nanjing University, Nanjing, China

Thulani Sibanda

Applied and Environmental Microbiology Research Group (AEMREG), Department of Biochemistry and Microbiology, University of Fort Hare, P/Bag X1314, Alice 5700, South Africa

Timothy Sibanda

Applied and Environmental Microbiology Research Group (AEMREG), Department of Biochemistry and Microbiology, University of Fort Hare, P Bag X1314, Alice, Eastern Cape, 5700, South Africa

Zoraida Sosa-Ferrera

Departamento de Química, Universidad de Las Palmas de Gran Canaria, Las Palmas de Gran Canaria, 35017, Spain

Rafael Szczepanowski

Institute for Genome Research and Systems Biology, Center for Biotechnology, Universität Bielefeld, Postfach 100131, D-33501 Bielefeld, Germany

Junying Tang

State Key Laboratory of Pollution Control and Resource Reuse, Environmental Health Research Center, School of the Environment, Nanjing University, Nanjing 210023, China

Zhu Wang

State Key Laboratory of Pollution Control and Resource Reuse, School of the Environment, Nanjing University, Nanjing, China

Jing Wang

National Institute of Metrology, Beijing 100013, China

Ke Xu

State Key Laboratory of Pollution Control and Resource Reuse, Environmental Health Research Center, School of the Environment, Nanjing University, Nanjing 210023, China

Min Yang
Research Center for Eco-Environmental Sciences, Chinese Academy of Sciences, Beijing 100085, China

Yu Zhang
Research Center for Eco-Environmental Sciences, Chinese Academy of Sciences, Beijing 100085, China

Xu-Xiang Zhang
State Key Laboratory of Pollution Control and Resource Reuse, School of the Environment, Nanjing University, Nanjing, China

Hong-Xun Zhang
College of Resources and Environment, Graduate University of the Chinese Academy of Sciences, Beijing 100049, China

INTRODUCTION

Inadequate wastewater treatment has serious consequences for human health. United Nations' statistics underline this: globally, 2 million tons of sewage, industrial, and agricultural waste are discharged into the world's waterways every year, while at least 1.8 million children under five years old die every year from water-related diseases. More people die as a result of polluted water than are killed by all forms of violence, including wars.

Meanwhile, cholera and other wastewater-related diseases are generally viewed as threats only for less developed countries. A recent report from the American Academy of Microbiology, however, indicates the developed world's complacency about wastewater treatment could be dangerous. The report estimates that, worldwide—including in developed nations—80 percent of infectious diseases may be water related. Diarrheal diseases traced to contaminated water cause about 900 million episodes of illness each year. Recent large-scale outbreaks of the waterborne illness cryptosporidiosis in the United Kingdom, Canada, and the United States illustrate the potential risk to developed countries.

Bacterial infection is not the only risk from inadequately treated wastewater. Pharmaceuticals in drinking water are another emerging issue. Antibiotics, antidepressants, and other medical and cosmetic pharmaceuticals are found in a large percentage of municipal water supplies. While these chemicals have the potential to impact public health, research suggests that their concentrations are usually low enough that they do not pose a major danger. A more serious concern, however, is the presence of antibiotics in wastewater. These select for resistance markers that are able to spread through the microbial community. As a result, antibiotic-resistant bacteria can potentially disseminate their resistance genes widely. The sludge products of urban and rural wastewater treatment plants are also increasingly used to fertilize agricultural crops, dispersing unknown amounts of resistance genes and antibiotics.

The articles selected for this compendium have been chosen to represent the most recent research into this worldwide problem. Such research has urgent relevance into today's world, both for policy-makers and for follow-up research.

—*Victor Monsalvo*

Human enteric viruses are causative agents in both developed and developing countries of many non-bacterial gastrointestinal tract infections, respiratory tract infections, conjunctivitis, hepatitis and other more serious infections with high morbidity and mortality in immunocompromised individuals such as meningitis, encephalitis, and paralysis. Human enteric viruses infect and replicate in the gastrointestinal tract of their hosts and are released in large quantities in the stools of infected individuals. The discharge of inadequately treated sewage effluents is the most common source of enteric viral pathogens in aquatic environments. Due to the lack of correlation between the inactivation rates of bacterial indicators and viral pathogens, human adenoviruses have been proposed as a suitable index for the effective indication of viral contaminants in aquatic environments. Chapter 1 reviews the major genera of pathogenic human enteric viruses, their pathogenicity and epidemiology, as well as the role of wastewater effluents in their transmission.

Vibrios are an example of an enteric pathogen that can be found in wastewater effluents of a healthy population. The authors of chapter 2 assess the prevalence of three non-cholerae vibrios in wastewater effluents of 14 wastewater treatment plants (WWTP) in Chris Hani and Amathole district municipalities in the Eastern Cape Province of South Africa for a period of 12 months. With the exception of WWTP10 where presumptive vibrios were not detected in summer and spring, presumptive vibrios were detected in all seasons in other WWTP effluents. When a sample of 1,000 presumptive *Vibrio* isolates taken from across all sampling sites were subjected to molecular confirmation for *Vibrio*, 668 were confirmed to belong to the genus *Vibrio*, giving a prevalence rate of 66.8%. Further, molecular characterisation of 300 confirmed *Vibrio* isolates revealed that 11.6% (35) were *Vibrio parahaemolyticus*, 28.6% (86) were *Vibrio fluvialis* and 28% (84) were *Vibrio vulnificus* while 31.8% (95) belonged to other *Vibrio* spp. not assayed for in this study. Antibiogram profiling of the three *Vibrio* species showed that *V. parahaemolyticus* was ≥50% susceptible to 8 of the test antibiotics and ≥50% resistant to only 5 of the 13 test antibiotics, while *V.*

vulnificus showed a susceptibility profile of ≥50 % to 7 of the test antibiotics and a resistance profile of ≥50% to 6 of the 13 test antibiotics. *V. fluvialis* showed ≥50 % resistance to 8 of the 13 antibiotics used while showing ≥50 % susceptibility to only 4 antibiotics used. All three *Vibrio* species were susceptible to gentamycin, cefuroxime, meropenem and imipenem. Multiple antibiotic resistance patterns were also evident especially against such antibiotics as tetracyclin, polymixin B, penicillin G, sulfamethazole and erythromycin against which all *Vibrio* species were resistant. These results indicate a significant threat to public health in a region of Africa that is characterized by widespread poverty, with more than a third of the population directly relying on surface water sources for drinking and daily use.

The aim of the authors of chapter 3 was to determine the occurrence of *Listeria* spp. in various points of a municipal wastewater treatment plant. The samples were collected from influent, effluent, raw sludge, stabilized sludge, and dried sludge from a wastewater treatment plant in Isfahan, Iran. The presence of *Listeria* spp. was determined using USDA procedures and enumerated by a three-tube most-probable-number assay using Fraser enrichment broth. Then, biochemically identified *Listeria monocytogenes* was further confirmed by PCR amplification. *L. monocytogenes*, *L. innocua*, and *L. seeligeri* were isolated from 76.9%, 23.1%, and 23.1% of influent; 38.5%, 46.2%, and 7.7% of effluent; 84.6%, 69.2%, and 46.2% of raw sludge; 69.2%, 76.9%, and 0% of stabilized sludge; and 46.2%, 7.7%, and 0% of dried sludge samples, respectively. The efficiency of wastewater treatment processes, digester tank, and drying bed in removal of *L. monocytogenes* were 69.6%, 64.7%, and 73.4%, respectively. All phenotypically identified *L. monocytogenes* were further confirmed by PCR method. The authors conclude that the application of sewage sludge in agricultural farms as fertilizer may result in bacteria spreading in agriculture fields and contaminated plant foods. This may cause a risk of spreading disease to human and animals. Using parameters such as BOD5 is not a sufficient standard for the elimination of pathogenic microorganisms.

In chapter 4, antimicrobial-resistance patterns were analyzed in *Escherichia coli* isolates from raw and treated wastewater (from two wastewater treatment plants, their marine outfalls, and the mouth of the Vistula River). The susceptibility of *E. coli* was tested against different classes of antibiotics. Isolates resistant to at least one antimicrobial agent were PCR tested for the presence of integrons. Ampicillin-resistant *E. coli* were the most frequent, followed by amoxicillin/clavulanate (up to 32%), trimethoprim/sulfamethoxazole (up to 20%), and fluoroquinolone (up to 15%)-resistant isolates. The presence of class 1 and 2 integrons was detected among tested *E. coli* isolates, with a rate of 32.06% ($n=84$) and 3.05% ($n=8$), respectively. The presence of integrons was associated with increased frequency of resistance to fluoroquinolones, trimethoprim/sulfamethoxazole, amoxicillin/clavulanate, piperacillin/tazobactam, and presence of multi-drug-resistance phenotype. Variable regions were detected in 48 class 1 and 5 class 2 integron-positive isolates. Nine different gene cassette arrays were confirmed among sequenced variable regions, with a predominance of *dfrA1-aadA1*, *dfrA17-aadA5*, and *aadA1* arrays. These findings illustrate the important role wastewater treatment plants play in spreading resistance genes in the environment. The author's findings also highlight the need for inclusion of monitoring efforts in the regular WWTP processes.

Antibiotic resistance represents a global health problem, requiring better understanding of the ecology of antibiotic resistance genes (ARGs), their selection, and their spread in the environment. Antibiotics are constantly released into the environment through wastewater treatment plant effluents. The authors of chapter 5 investigated the effect of these discharges on the prevalence of ARGs and bacterial community composition in biofilm and sediment samples of a receiving river. They used culture-independent approaches such as quantitative PCR to determine the prevalence of eleven ARGs, and they used 16S rRNA gene-based pyrosequencing to examine the composition of bacterial communities. Concentration of antibiotics in WWTP influent and effluent were also determined. ARGs such as *qnrS*, *bla*$_{TEM}$, *bla*$_{CTX-M}$, *bla*$_{SHV}$, *erm*(B), *sul*(I), *sul*(II), *tet*(O), and *tet*(W) were detected in all biofilm and sediment samples analyzed. Moreover, the authors observed a significant increase in the relative abundance of ARGs in biofilm samples collected downstream of the WWTP discharge. They also found significant differences with respect to community structure and composition between upstream and downstream samples. Their results indicate that WWTP discharges may contribute to the spread of ARGs into the environment and may also impact on the bacterial communities of the receiving river.

The reuse of treated wastewater (TWW) for irrigation is a practical solution for overcoming water scarcity, especially in arid and semiarid regions of the world. However, there are several potential environmental and health-related risks associated with this practice. One such risk comes from the fact that TWW irrigation may increase antibiotic resistance (AR) levels in soil bacteria, potentially contributing to the global propagation of clinical AR. Wastewater treatment plant effluents have been recognized as significant environmental AR reservoirs due to selective pressure generated by antibiotics and other compounds that are frequently detected in effluents. The review in chapter 6 summarizes many recent studies that have assessed the impact of anthropogenic practices on AR in environmental bacterial communities, with specific emphasis on elucidating the potential effects of TWW irrigation on AR in the soil microbiome. Based on the current state of the art, the authors conclude that contradictory to freshwater environments where WWTP effluent influx tends to expand antibiotic-resistant bacteria (ARB) and antibiotic-resistant genes levels, TWW irrigation does not seem to impact AR levels in the soil microbiome. Although this conclusion is a cause for cautious optimism regarding the future implementation of TWW irrigation, the authors also conclude that further studies aimed at assessing the scope of horizontal gene transfer between effluent-associated ARB and soil bacteria need to be further conducted before ruling out the possible contribution of TWW irrigation to antibiotic-resistant reservoirs in irrigated soils.

Antibiotics are often used to prevent sickness and improve production in animal agriculture, which means their residues in animal bodies may enter tannery wastewater during leather production. The study found in chapter 7 used Illumina high-throughput sequencing to investigate the occurrence, diversity, and abundance of antibiotic resistance genes and mobile genetic elements (MGEs) in aerobic and anaerobic sludge of a full-scale tannery wastewater treatment plant. Metagenomic analysis showed that *Proteobacteria, Firmicutes, Bacteroidetes* and *Actinobacteria* dominated in the WWTP, but the relative abundance of archaea in anaerobic sludge was higher than in aerobic sludge. Sequencing reads from aerobic and anaerobic sludge revealed differences in the abundance of functional genes between both microbial communities. Genes coding for antibiotic resistance were identified in both communities. BLAST analysis against the Antibiotic Resistance Genes Database further revealed that aerobic and anaerobic sludge contained various ARGs with high abundance, among which sulfonamide resistance gene *sul1* had the highest abundance, occupying over 20% of the total ARGs reads. Tetracycline resistance genes (*tet*) were highly rich in the anaerobic sludge, among which *tet33* had the highest abundance, but was absent in aerobic sludge. Over 70 types of insertion sequences were detected in each sludge sample, and class 1 integrase genes were prevalent in the WWTP. The results highlighted prevalence of ARGs and MGEs in tannery WWTPs, indicating they deserve more attention from those who are responsible for protecting public health.

In order to comprehensively investigate tetracycline resistance in activated sludge of sewage treatment plants, the authors of chapter 8 used 454 pyrosequencing and Illumina high-throughput sequencing to detect potential tetracycline resistant bacteria (TRB) and antibiotic resistance genes in sludge cultured with different concentrations of tetracycline. Pyrosequencing of 16S rRNA gene revealed that tetracycline treatment greatly affected the bacterial community structure of the sludge. Nine genera consisting of Sulfuritalea, Armatimonas, Prosthecobacter, Hyphomicrobium, Azonexus, Longilinea, Paracoccus, Novosphingobium and Rhodobacter were identified as potential TRB in the sludge. Results of qPCR, molecular cloning, and metagenomic analysis consistently indicated that tetracycline treatment could increase both the abundance and diversity of the tet genes, but decreased the occurrence and diversity of non-tetracycline ARG, especially sulfonamide resistance gene sul2. Cluster analysis showed that tetracycline treatment at subinhibitory concentrations (5 mg/L) was found to pose greater effects on the bacterial community composition, which may be responsible for the variations of the ARGs abundance. This study indicated that joint use of 454 pyrosequencing and Illumina high-throughput sequencing can be effectively used to explore ARB and ARGs in the environment, and future studies should include an in-depth investigation of the relationship between microbial community, ARGs and antibiotics in sewage treatment plant (STP) sludge.

To detect plasmid-borne antibiotic-resistance genes in wastewater treatment plant bacteria, the authors of chapter 9 designed and synthesized 192 resistance-gene-specific PCR primer pairs. Subsequent PCR analyses on total plasmid DNA preparations obtained from bacteria of activated sludge or the WWTP's final effluents led to the identification of, respectively, 140 and 123 different resistance-gene-specific amplicons. The genes detected included aminoglycoside, β-lactam, chloramphenicol, fluoroquinolone, macrolide, rifampicin, tetracycline, trimethoprim and sulfonamide resistance genes as well as multidrug efflux and small multidrug resistance genes. Some of these genes were only recently described from clinical isolates, demonstrating genetic exchange between clinical and WWTP bacteria. Sequencing of selected resistance-gene-specific amplicons confirmed their identity or revealed that the amplicon nucleotide sequence is very similar to a gene closely related to the reference gene used for primer design. These results demonstrate that WWTP bacteria are a reservoir for various resistance genes. Moreover, detection of about 64% of the 192 reference resistance genes in bacteria obtained from the WWTP's final effluents indicates that these resistance determinants might be further disseminated in habitats downstream of the sewage plant.

Antidepressants are a new kind of pollutant that is increasingly found in wastewater. In chapter 10, a fast and sensitive ultra-high performance liquid chromatography-tandem mass spectrometry method (UHPLC-MS/MS) was developed and validated for the analysis of 24 antidepressant drugs and six of their metabolites in wastewater. This is the first time that the antidepressant residues in wastewater of Beijing (China) were systematically reported. A solid-phase extraction process was performed with 3 M cation disk, followed by ultra-high performance liquid chromatography–tandem mass spectrometry measurements. The chromatographic separation and mass parameters were optimized in order to achieve suitable retention time and good resolution for analytes. All compounds were satisfactorily determined in one single injection within 20 min. The limit of quantification (LOQ), linearity, and extraction recovery were validated. The LOQ for analytes were ranged from 0.02 to 0.51 ng/mL. The determination coefficients were more than 0.99 within the tested concentration range (0.1–25 ng/mL), and the recovery rate for each target compound was ranged from 81.2% to 118% at 1 ng/mL. At least ten target antidepressants were found in all samples, and the highest mean concentration of desmethylvenlafaxin was up to 415.6 ng/L.

Since trace organic compounds such as pharmaceuticals in surface water have been a relevant threat to drinking water supplies, in the study described in chapter 11, removal of pharmaceuticals and transformation of pharmaceuticals into metabolites were investigated in the main sources of micropollutants, such as WWTPs and engineered constructed wetlands. Pharmaceuticals were effectively removed by different WWTP processes and wetlands. Pharmaceutical metabolites with relatively low log D value were resulted in the low removal efficiencies compared to parent compounds with relatively high log D value, indicating the stability of metabolites. The constructed wetlands fed with wastewater effluent were encouraged to prevent direct release of micropollutants into surface waters. Among various pharmaceuticals, different transformation patterns of ibuprofen were observed with significant formation of 1-hydroxy-ibuprofen during biological treatment in WWTP, indicating preferential biotransformation of ibuprofen. Lastly, transformation of pharmaceuticals depending on their structural position was investigated in terms of electron density, and, the electron rich $C_1 = C_2$ bond of carbamazepine was revealed as an initial transformation position.

In chapter 12, an assessment of the concentrations of thirteen different therapeutic pharmaceutical compounds was conducted on water samples obtained from different wastewater treatment plants. The authors used solid phase extraction, and high- and ultra-high-performance liquid chromatography with mass spectrometry detection (HPLC-MS/MS and UHPLC-MS/MS) was carried out. The target compounds included ketoprofen and naproxen (anti-inflammatories), bezafibrate (lipid-regulating), carbamazepine (anticonvulsant), metamizole (analgesic), atenolol (β-blocker), paraxanthine (stimulant), fluoxetine (antidepressant), and levofloxacin, norfloxacin, ciprofloxacin, enrofloxacin and sarafloxacin (fluoroquinolone antibiotics). The relative standard deviations obtained by this method were below 11%, while the detection and quantification limits were in the range of $0.3 - 97.4$ ng/L^{-1} and $1.1 - 324.7$ ng/L^{-1}, respectively. The water samples were collected from two different WWTPs located on the island of Gran Canaria in Spain over a period of one year. The first WWTP used conventional activated sludge for the treatment of wastewater, while the other plant employed a membrane bioreactor system for wastewater treatment. Most of the pharmaceutical compounds detected in this study during the sampling periods were found to have concentrations ranging between 0.02 and 34.81 μg/L^{-1}.

PART I

DISEASE EXPOSURE

CHAPTER 1

Inadequately Treated Wastewater as a Source of Human Enteric Viruses in the Environment

ANTHONY I. OKOH, THULANI SIBANDA, AND SIYABULELA S. GUSHA

1.1 INTRODUCTION

Human enteric viruses are obligate parasites of man that infect and replicate in the gastrointestinal tract of their hosts. Patients suffering from viral gastroenteritis or viral hepatitis may excrete about 105 to 1011 virus particles per gram of stool [1], comprising various genera such as adenoviruses, astroviruses, noroviruses, Hepatitis E virus, parvoviruses, enteroviruses (Coxsackie viruses, echoviruses and polioviruses), Hepatitis A virus, and the rotaviruses [2]. Consequently virus concentrations in raw water receiving fecal matter are often high; although viruses cannot reproduce in water they are still capable of causing diseases when ingested, even at low doses [3].

Human enteric viruses are causative agents of many non-bacterial gastrointestinal tract infections, respiratory infections, conjunctivitis, hepatitis and other serious infections such as meningitis, encephalitis and paralysis. These are common in immunocompromised individuals with high morbidity and mortality attributable to these infections in both developed and developing countries. Most cases of enteric virus infections have particularly been observed to originate from contaminated drinking water sources, recreational waters and foods contaminated by sewage and sewage effluents waters [4].

Wastewater treatment processes such as the activated sludge process, oxidation ponds, activated carbon treatment, filtration, and lime coagulation and chlorination only eliminate between 50% and 90% of viruses present in wastewater [5], allowing for a significant viral load to be released in effluent discharge. Due to their stability and persistence, enteric viruses subsequently become pollutants in environmental waters resulting in human exposure through pollution of drinking water sources and recreational waters, as well as foods. The performance of wastewater treatment systems is at present monitored largely by the use of bacterial indicator organisms. Considering that infectious viruses have been isolated from aquatic environments meeting bacterial indicator standards, in some instances in connection with virus related outbreaks [6], the use of bacterial indicators has thus been considered an insufficient tool to monitor wastewater quality because bacterial and viral contaminations are not necessarily associated and linked with each other [7]. This paper reviews the major genera of pathogenic human enteric viruses, their pathogenicities and epidemiology, as well as the role of wastewater effluents in their transmission.

1.2 MAJOR GENERA OF HUMAN ENTERIC VIRUSES: STRUCTURE, PATHOGENICITY AND EPIDEMIOLOGY

A diverse range of enteric virus genera and species colonize the gastrointestinal tracts of humans producing a range of clinical manifestations and varying epidemiological features. From a public health perspective, the most important of these are the rotaviruses, adenoviruses, noroviruses, enteroviruses as well as Hepatitis A and E viruses.

1.2.1 ROTAVIRUSES

Rotaviruses are large 70 nm nonenveloped icosahedral viruses that belong to the family Reoviridae [8]. A rotavirus particle consists of a triple-layered protein capsid enclosing 11 segments of a double-stranded RNA genome [9]. The genome encodes six viral proteins (VP1, VP2, VP3, VP4, VP6 and VP7) that make up the viral capsid, and five non-structural proteins (NSP1–NSP5) [10]. The outer capsid is primarily composed of VP4 (a protease-sensitive protein designated P) and VP7 (a glycoprotein designated G) which also forms the basis of defining rotaviruses into P and G serotypes [8]. These two proteins are also determinants of host range. In particular VP4 has been shown to be a determinant of several important functions, such as cell attachment, entry into cells, hemagglutination, and neutralization [9].

There are seven species of rotaviruses, designated A to G, of which groups A–C infect humans [11]. At least 14 G types (G1 to G14) and 20 P types (P [1] to P [20]) have been identified to date, of which 10 G types and five P types have been found in rotaviruses infecting humans [12]. The occurrence of these strains varies spatially and temporally. Type G1P [8] strains are unanimously regarded as the most prevalent and ubiquitous while types G2P [4], G3P [8], and G4P [8] are ubiquitous, but their diffusion is temporal and regional [13].

Rotaviruses infect mature enterocytes in the mid and upper villous epithelium of the host's small intestines [14]. During the rotavirus replication cycle, virions attach to host cells as triple-layered particles and subsequently enter the cytoplasm by either plasma membrane or endosomal membrane penetration. The attachment of the virus to the cells of the intestinal mucosa is mediated by the structural protein VP4. The infectivity of the virus is enhanced by cleavage of VP4 to produce VP8* and VP5*. The binding of the virus has been proposed to be initially mediated by the cleavage protein VP8* through N-acetylneuraminic (sialic) acid residues on the cell surface membrane of the host cell, followed by VP5* or directly by VP5* without the involvement of sialic acid residues. In both cases, the identity of the receptors has remained unclear although, they are thought to be part of lipid micro domains [15].

As a result of cell entry, the outer layer of VP4/VP7 is lost, and the resulting double-layered particles become transcriptionally active, releasing mRNA transcripts through a system of channels that penetrate the middle (VP6) and inner (VP2) capsid layers at each of the icosahedral vertices [16]. After cytosolytic replication in the mature enterocytes of the small intestine, new rotavirus particles can infect distal portions of the small intestine or be excreted in the feces [17].

The pathology of rotavirus infections have been based on a few studies of the jejunal mucosa of infected infants which have revealed shortening and atrophy of villi, distended endoplasmic reticulum, mononuclear cell infiltration, mitochondrial swelling and denudation of microvilli [15]. Rotavirus infection alters the function of the small intestinal epithelium, resulting in the destruction of the mature enterocytes that are responsible for the absorptive function of the villi, while favouring the proliferation of crypt cells that are more secretory resulting in malabsorptive diarrhea [18]. The decreased absorption of Na+ ions, results in the transit of undigested mono- and disaccharides, fats, and proteins into the colon. The undigested bolus is osmotically active, resulting in impairment water absorption by the colon which leads to an osmotic diarrhea [18]. The classic presentation of rotaviral infection is fever and vomiting for 2–3 days, followed by non-bloody diarrhea. The diarrhea may be profuse, and 10–20 bowel movements per day are common. When examined, the stool from infected patients is generally devoid of fecal leucocytes [17]. Severe rotavirus gastroenteritis has been associated with pancreatitis [19].

Rotaviruses have been recognized as the leading cause of severe diarrhea in children below 5 years of age, with an estimated 140 million cases and about 800,000 deaths and about 25% of all diarrheal hospital admissions in developing countries each year [20]. Group A rotaviruses are the species most frequently associated with acute gastroenteritis in developed and developing countries. At present there is no available specific treatment for rotavirus infection [21], except prevention through vaccination that has gained licensing in many developed and developing countries [22].

1.2.2 ENTEROVIRUSES

Human enteroviruses are members of the family Picornaviridae, which consist of nonenveloped virus particles containing a 7,500-nucleotide single-stranded positive sense RNA genome protected by an icosahedral capsid [23]. The genome encodes four structural proteins, VP1 to VP4 and seven nonstructural proteins implicated in viral replication and maturation. The capsid proteins VP1, VP2, and VP3 are located at the surface of the capsid and are therefore containing epitopes for immunological reaction [23]. There are more than 80 serotypes of human enteroviruses that have been identified on the basis of traditional neutralization tests which aided by the use of molecular based techniques like nucleic acid sequencing has revealed new strains [24]. On the basis of phylogenetic analysis of multiple genome regions, the enterovirus serotypes are classified into four species (Human enterovirus A-D) [25]. These groups consist of 31 serotypes of Echovirus, 23 serotypes of Coxsackie A virus, six serotypes of Coxsackie B virus, three serotypes of Poliovirus and the numbered serotypes of enterovirus [26].

The pathogenicity of enteroviruses is mediated by an arginine-glycine-aspartic acid (RGD) motif found on the viral capsid proteins of the picornavirus family [27]. About seven distinct receptors for different enteroviruses have been identified from human cells, namely; the poliovirus receptor (PVR; CD155), three integrins ($\alpha2\beta1$, $\alpha v\beta3$, and $\alpha v\beta6$), decay-accelerating factor (DAF; CD55), the coxsackievirus-adenovirus receptor (CAR), and intracellular adhesion molecule 1 (ICAM-1) [26]. Typically, the primary site of infection is the epithelial cells of the respiratory or gastrointestinal tract. From the primary infection site, the viruses may spread to secondary sites particularly following viremia. Secondary infection of the central nervous system results in aseptic meningitis or, rarely, encephalitis or paralysis [26].

Most enterovirus infections are asymptomatic or result in only mild illnesses, such as non-specific febrile illness or mild upper respiratory tract infections. However, enteroviruses can also cause a wide variety of clinical illnesses including acute haemorrhagic conjunctivitis, aseptic meningitis, undifferentiated rash, acute flaccid paralysis, myocarditis and neonatal sepsis-like disease [28]. Enteroviruses are the most common etiological agents of human viral myocarditis and are associated with some cases of dilated cardiomyopathy (DCM), which alone afflicts approximately five to eight persons per 100,000 per year worldwide [29]. Enteroviruses are cytopathic, most infections result in tissue specific cell destruction, although some disease manifestations can be a result of host immune response [26].

One of the most distinctive enterovirus diseases is poliomyelitis. It is almost invariably caused by one of the three poliovirus serotypes. Polioviruses may also cause aseptic meningitis or nonspecific minor illness [30]. The normal route of poliovirus infection in naturally permissive hosts begins with infection of the enteric system through oral ingestion of the virus [31]. The cell receptor for all three poliovirus serotypes is CD155, a glycoprotein that is a member of the immunoglobulin super family of proteins [32]. Viral particles initially replicate in the gastrointestinal system, but replication at this site does not result in any detectable pathology [31]. From the primary sites of multiplication in the mucosa, the virus drains into cervical and mesenteric lymph nodes and then to the blood, causing a transient viremia. Most natural infections of humans end at this stage with a minor disease comprising nonspecific symptoms such as sore throat, fever, and malaise. Replication at extraneural sites is believed to maintain viremia beyond the first stage and increase the likelihood of virus entry into the central nervous system. Such extraneural sites might include brown fat, reticuloendothelial tissues, and muscle [32]. As viremia spreads, the infection of dendritic cells and macrophages can aid the transport of the viruses across the blood-brain barrier or transport along neural pathways to infect brain cells [31].

1.2.3 ADENOVIRUSES

Adenoviruses are nonenveloped viruses, about 90 nm in diameter with a linear, double-stranded DNA genome of 34–48 kb and an icosahedral capsid [33]. On the basis of hemagglutination properties as well as DNA sequence homology, tissue tropism, fiber protein characteristics, and other biological properties, human adenoviruses are classified into six species designated A to F [34]. The six species consist of 51 serotypes, defined mainly by neutralization criteria [10]. The virus capsid contains at least nine proteins, of which the hexon, penton base and the fibre proteins are the major capsid proteins [33]. The penton base and the elongated fiber protein form a complex at the vertex of the virus capsid [35].

Adenovirus infection of host epithelial (gastrointestinal and respiratory) cells is mediated by the fibre and penton base capsid proteins. In the case of adenovirus subgroups A and C–F, the attachment to cells is mediated by a high affinity binding of the fiber protein to a 46 kDa membrane protein known as the coxsackie adenovirus receptor (CAR), a member of the immunoglobulin receptor super family serving as a cell to cell adhesion molecule in tight junctions [36]. Subgroup B serotypes such as Ad3, Ad11, and Ad33, as well as the subgroup D serotype Ad37 utilize other receptors such as CD46 and sialic acid [36]. The entry and internalization of the virus into host cells is facilitated by the penton base through the binding of the conserved arginine-glutamine-aspartic acid (RGD) motif to $\alpha v\beta3$ or $\alpha v\beta5$ integrins leading to endocytosis [35].

The major receptor for adenoviruses, CAR is not normally accessible from the apical surfaces. As a result, the initial adenovirus infection is presumed to occur through transient breaks in the epithelium allowing the luminal virus to reach its receptor or during the repair of injured epithelium when CAR might be accessible [37]. Following viral replication, infected cells release viral particles which then filter through the leaky paracellular pathway to emerge on the apical surface where they can spread to other sites of infected tissues [38]. The adenovirus fiber-CAR interactions are also thought to play a role in systemic spread of the virus through the disruption of CAR-mediated endothelial cell-cell adhesion which could facilitate spread to the bloodstream, and virus transport to other sites in the body [38]. The viral infection of the respiratory or gastrointestinal tract may lead to widespread dissemination which can result in diseases such as pharyngitis, conjunctivitis, pneumonia, haemorrhagic cystitis, colitis, hepatitis, or encephalitis which may be fatal in children and immunocompromised patients [39].

Adenovirus infections occur worldwide throughout the year [40]. The serotypes most frequently associated with respiratory infection are members of the subgroup B (Ad3, Ad7, and Ad21), species C (Ad1, Ad2, Ad5 and Ad6) and species E (Ad4) [34].

1.2.4 NOROVIRUSES

Noroviruses are members of the family Caliciviridae [2]. Noroviruses contain a single-stranded positive sense RNA genome of approximately 7.7 kb which is organized into three open reading frames (ORFs). ORF1 encoding a 200-kDa polyprotein that is processed into at least six non-structural proteins; ORF2 encodes a 60-kDa capsid protein VP1 and ORF3 encoding a basic minor structural protein VP2 [41]. The exterior surface of the virion is composed of a single major protein VP1 that forms the capsid and appears as 32 cup-shaped depressions on the surface showing an icosahedral symmetry on microscopy [42].

The VP1 subunit consists of a shell (S) and a protruding (P) domain that is made up of a middle P1 and a distal P2 subdomains [43]. While the S domain is responsible for the icosahedral shell structure, the P1 and P2 subdomains have been implicated in antigenicity and cellular receptor binding of these viruses [41]. Binding of the VP1 proteins occurs through human histoblood group antigens (HBGAs) as receptors. Human HBGAs are present on the surfaces of red blood cells and more importantly, on the mucosal epithelium [44].

Noroviruses are a major cause of acute viral gastroenteritis, affecting people of all age groups worldwide [45]. Outbreaks of norovirus gastro-enteritis can be seasonal or sporadic cases that occur through out the year [46] especially in semiclosed communities such as families, schools, elderly people's homes, hospitals, hotels, and cruise ships [47].

1.3 THE WASTEWATER TREATMENT PROCESS AND POLLUTION FROM VIRAL PATHOGENS

Municipal wastewater is a mixture of human excreta (sewage), suspended solids, debris and a variety of chemicals that originate from residential, commercial and industrial activities [48]. Raw sewage is a major carrier of disease causing agents, particularly enteric pathogens [1]. The safe treatment of sewage is thus crucial to the health of any community. In subjecting municipal wastewater to treatment before discharge to the environment, the goal is to remove pollutants, both chemical and biological, from the water in order to decrease the possibility of detrimental impacts on humans and the rest of the ecosystem [49]. In the conventional municipal wastewater treatment systems, physical processes such as sedimentation, activated sludge and trickling filters are often used in the decontamination of the wastewater. Human enteric viruses exist in waters as either free-floating or adsorbed onto solid particles. Physical removal of particles by processes like coagulation, flocculation, sedimentation and filtration aids the removal of viruses in wastewater effluents [50]. While these processes remove some viruses associated with large particles, smaller colloidal particles (<10 μm) may pass through these processes to the disinfection stages where they continue to enmesh and protect viruses against disinfectant action [51]. These physical processes remove about 90–99% of the viral load of the wastewater [52]. Additional removal of biological pollutants is achieved by disinfection which often uses chlorine and sometimes ozone, paracetic acid and UV irradiation [53]. Although the combination of all these processes may remove a substantial load of viruses, their efficiencies may vary leading to discharge of pathogenic viruses in the effluents where they subsequently become environmental pollutants [54].

The assessment of the microbiological quality of wastewater effluents has traditionally depended on indicator organisms, such as coliforms or enterococci, which however do not always reflect the risk of other microbial pathogens such as viruses, stressed bacterial pathogens and protozoa [55]. In particular the indicator bacteria survival in water does not correlate with that of enteric viruses [56]. Our recent studies [57–59] have shown that the wastewater treatment facilities in the Eastern Cape Province of South Africa are a veritable source of pathogens in aquatic environments of this study area and negatively impact physico-chemical quality of receiving watershed [60], therefore it is highly probable that they might also be a source of enteric viruses in the aquatic environment.

Viral pathogens have frequently been detected in waters that comply with bacterial standards [61,62]. Human enteric viruses as gastrointestinal tract pathogens are shed in large quantities in the fecal waste of infected individuals and are therefore also found in high quantities in raw sewage [63]. The extent of enteric virus reduction varies according to the sewage treatment system used and the virus type [64].

1.4 FACTORS AFFECTING THE REMOVAL AND INACTIVATION OF VIRUSES IN WASTEWATER SYSTEMS

Enteric viruses in wastewater treatment plants are removed by a combination of irreversible adsorption as well as inactivation by disinfectants [65]. Processes such as coagulation, flocculation, sedimentation and filtration remove viruses adsorbed onto particulate matter [66,67]. The efficiency of removal varies depending on the adsorptive affinities of the virus particles and the adsorbents [68]. Potential adsorbents of viruses in natural waters include sand, pure clays (e.g., montmorillonite, illite, kaolinite, and bentonite), bacterial cells, naturally occurring suspended colloids, and estuarine silts and sediments [50]. Removal rates depend to a great extent on the pH, substrate saturation, redox potential and dissolved oxygen of the system. The protein coats of most viruses gives the viral particles a net charge due to the presence of amino acids such as glutamic acid, aspartic acid, histidine and tyrosine that contain ionized carboxylic and amino groups. Most enteric viruses have a net negative charge at a pH above 5 and a net positive charge below pH 5 [69]. The adsorptive interaction between the virus particle and the adsorbents is a function of isoelectric point of the virus, as well as that of the adsorbent particle and also its hydrophobicity. The variation of dissociation constants among the various polypeptides ensures that most viruses have net charges that vary continuously with varying pH [50]. Adsorption may also be affected by factors such as flow rate and ionic strength. Also flow rate may affect the contact of viruses' attachment sites, with increasing velocities reducing contact time and therefore the subsequent attachment to sediments. High ionic strength, such as septic tank effluent, favour virus adsorption, with low ionic strength waters, such as rainfall, able to remobilize attached viruses [65].

The inactivation of viruses by disinfection is a process affected by suspended particles. Disinfection relies on the ability of either chemical disinfectant molecules or high-energy photons (in the case of UV disinfection) coming into contact with the viruses [50]. Chemical disinfectants inactivate viruses by either oxidation or disintegration of viral particle, or inhibition of cellular activity [70]. UV disinfection on the other hand relies on the formation of pyrimidine dimers in the DNA/RNA of the target organism, which prevents replication [71]. If contact between the disinfecting agent and the organism is reduced or prevented altogether, then disinfection may be impeded [50]. Organic particles negatively impact the chemical disinfection of viruses by creating a demand for the disinfectant molecules as they penetrate the particle surface. In addition to the disinfectant demand of the particle, particle structure and porosity also plays a role in the shielding of viruses from disinfection [72]. The presence of particle-associated viruses during disinfection of water results in reduced virus inactivation compared to particle-free waters [73].

1.5 RESISTANCE OF ENTERIC VIRUSES TO DISINFECTANTS

The study of the inactivation of enteric viruses following wastewater disinfection is complicated by the low and variable levels of enteric viruses frequently seen in effluents [74]. Research has demonstrated that enteric viruses are inherently more resistant to common disinfectants than bacterial indicators. Tree et al. [74], observed that bacterial indicators *Escherichia coli* and *Enterococcus faecalis* were rapidly inactivated by chlorine with inactivation levels of (>5 log10 units) while there was poor inactivation (0.2 to 1.0 log10 unit) of F+-specific RNA (FRNA) bacteriophage (MS2) at doses of 8, 16, and 30 mg/liter of free chlorine. Armon et al. [75] also showed that the inactivation levels of naturally occurring coliphages were significantly lower than that of coliforms after chlorination. With regards to UV radiation, enteric adenoviruses have also been shown to be more resistant than bacterial spores [76].

In the United States, the Environmental Protection Agency (EPA) recommends the use of an additional criterion for the evaluation of water disinfection based on viral inactivation. The standard makes use of Ct values, defined as disinfectant concentration (C) multiplied by the contact time (t) between the disinfectant and microorganism. The recommendations direct that public utilities must ensure a 4-log (Ct 99.99%) inactivation of viruses [77].

1.6 CONSEQUENCES OF ENTERIC VIRUS PERSISTENCE IN WASTEWATER EFFLUENTS

The inability of wastewater treatment systems to ensure a complete in-activation of viruses in wastewater effluents has serious implications on public health. Virus levels in treated wastewater, measured by cell culture assay, range from 1.0×10^{-3} to 1.0×10^{2} liter^{-1} depending on the level of treatment [78]. Human enteric viruses can remain stable in the environment for long periods particularly in association with solids in sediments. Goyal et al. [79] detected human enteric viruses in sedi-ments obtained from sewage sludge disposal sites in the Atlantic Ocean 17 months after the cessation of sludge dumping. The sediments act as a reservoir from which viruses are resuspended in the water [1]. The persistence of enteric viruses in environmental waters often leads to in-cidences of human infection through contamination of food, drinking and recreational waters. Enteric viruses have very low infectious doses in the order of tens to hundreds of virions [80]. Even high log reductions in concentration during transport could still result in infectious viruses present in potable water or food [80].

1.7 WATER

The discharge of inadequately treated sewage water has a direct impact on the microbiological quality of surface waters and consequently the potable water derived from it. The inherent resistance of enteric viruses to water disinfection processes means that they may likely be present in drinking water exposing consumers to the likelihood of infection. In one study, Human adenoviruses were detected in about 22% of river water samples and about 6% of treated water samples in South Africa [81]. In another study, about 29% of river water samples and 19% of treated drinking water samples in South Africa had detectable levels of entero-viruses [61].

Enteric viruses are the most likely human pathogens to contaminate groundwater. Their extremely small size, allows them to infiltrate soils from contamination sources such as broken sewage pipes and septic tanks, eventually reaching aquifers. Viruses can move considerable distances in the subsurface environment with penetration as great as 67 m and horizontal migration as far as 408 m [80]. In a study in the United States, 72% of groundwater sites were positive for human enteric viruses [82]. In America, the U.S. Environmental Protection Agency (EPA) has proposed a Groundwater Rule that requires public groundwater sites considered to be vulnerable to fecal pollution to be monitored monthly for fecal indicators and that where indicators are found, they must either be a removal of pollution sources or disinfection [82]. Groundwater has been implicated as a common transmission route for waterborne infectious disease in the United States with about 80% waterborne outbreaks attributed to drinking contaminated well water. The enteric viruses most frequently associated with outbreaks are noroviruses and hepatitis A virus [80].

Another important human exposure pathway is through recreational waters. Human enteric viruses have frequently been detected in coastal waters receiving treated wastewater effluents. Xagoraraki et al. [83] reported human adenovirus concentrations at the level of 10^3 virus particles·liter^{-1} in recreational beaches in America. Mocé-Llivina et al. [84] detected enteroviruses in 55% of samples from beaches in Spain. The occurrence of viruses in coastal waters results in increased risks of infection to swimmers and divers. The risk of ear, eye, gastrointestinal or respiratory infections is more than twice in polluted than unpolluted beaches [81].

Numerous outbreaks of enteric virus associated diarrhea have been linked to the consumption of water contaminated with viruses. Kukkula et al. [85], showed a strong epidemiological risk ratio between the consumption of water contaminated with noroviruses and the outbreak of acute gastroenteritis in Finland. Karmakar et al. [6] reported a water-borne outbreak of rotavirus gastroenteritis in India.

1.8 CONTAMINATION RISKS OF FOODS
FROM WASTEWATERS WITH POLLUTANT ENTERIC VIRUSES

Viral contaminants may persist on food surfaces or within foods for extended periods [86]. Pre-harvest contamination may occur in agricultural products subjected to irrigation with reclaimed wastewater, crop fertilization with sewage sludge, or fecal pollution of the areas in which food products are obtained. Numerous studies have attributed outbreaks of enteric virus diseases such as acute gastroenteritis and hepatitis A to the consumption of raw vegetables such as salads. Using epidemiologic data in a case controlled study, Grotto et al. [87] showed an association between a norovirus outbreak of gastroenteritis at a military camp in Israel and the consumption of vegetable salads 48 hours preceding the outbreak. In Sweden, Le Guyader et al. [86] using sequence based molecular fingerprinting also reported that acute gastroenteritis outbreak was a result of consumption of raspberry cakes contaminated with noroviruses.

Post-harvest contamination of raw food may occur as a result of human handling by workers and consumers, contaminated harvesting equipment, transport containers, contaminated aerosols, wash and rinse water or cross contamination during transportation and storage [88]. Recontamination after cooking or processing, and inadequate sanitation has also been associated with outbreaks of enteric virus infections [89]. In an outbreak of acute viral hepatitis A in Italy, Chironna et al. [90] using sequence-based molecular fingerprinting identified a point source of the virus outbreak as a food handler working at a local food outlet. A number of studies have also implicated enteric viruses in disease outbreaks involving contaminated foods [91].

Probably one of the most recognized food borne transmission of enteric virus infections is through the consumption of shellfish grown in sewage polluted marine environments. Shellfish, which includes molluscs such as oysters, mussels, cockles, clams and crustaceans such as crabs, shrimps, and prawns [92], are filter-feeders that result in the bio-concentration of environmentally stable, positive-stranded RNA viruses, such as norovirus, hepatitis A virus and enterovirus in their digestive glands and gills [93]. The risk of human exposure to enteric viral pathogen is increased by the fact that shellfish are often consumed raw, or only slightly cooked [94]. The consumption of shellfish growing in aquatic environments impacted by wastewater effluents or untreated sewage has been associated with numerous outbreaks of gastroenteritis caused by noroviruses as well as cases of hepatitis A [95]. Also, Karamoko et al. [96] report mussel samples positive for enteroviruses, and strongly suggests a connection between contaminations of foods by wastewater borne enteric viruses since these mussels were harvested from an area close to a domestic wastewater outlet, more so as mussels harvested from an aquaculture were all found not to be positive for enterovirus. In a similar report, a serious food-borne outbreak in China in 1988 [97–100] was attributed to consumption of clams contaminated with hepatitis A virus from a sewage-polluted community near Shanghai. In Israeli, it was demonstrated that communities using wastewater effluents for irrigation have high incidences of infectious hepatitis as compared to other communities [101]. Although there is little data on the role of wastewater effluents in the propagation of food borne viral diseases, there is high probability that this can be a significant mode of contamination and subsequent disease transmission [101].

1.9 FUTURE DIRECTIONS

Current safety standards for determining food and water quality typically do not specify what level of viruses should be considered acceptable. This is in spite of the fact that viruses are generally more stable than common bacterial indicators in the environment. While there has been a significant amount of research on the impacts of inadequately treated wastewater effluents in developed countries, the same can not be said of developing countries which coincidentally are faced with a huge burden of infectious diseases emanating from pollution of water bodies with wastewater effluent discharges (von Sperling and Chernicharo [102] most of which remains undocumented, unreported and not properly investigated. The major limitation has been the high cost of establishing facilities for the monitoring and surveillance, especially with enteric viruses that requires specialized laboratories and techniques such as tissue culture, electron microscopy and immunological assays. The use of molecular techniques such as PCR which are relatively rapid and specific however may prove useful for the monitoring of enteric viruses in wastewater effluents. This will have significant benefits in identifying potential avenues of transmission of infectious viruses.

The challenge in ensuring safe water with regards to viral pathogens is that the detection of putative indicators of viral pathogens such as bacteriophages does not always correlate with that of other viruses particularly pathogenic enteric viruses [103]. Human adenoviruses have been proposed as a suitable index for the effective indication of viral contaminants of human origin. For one reason, they are prevalent and very stable; for another, they are considered human specific and are not detected in animal wastewaters or slaughterhouse sewage [103]. Adenovirus strains Adv40 and Adv41 have been associated with diarrheal diseases which can be attributed to consumption of fecal contaminated water and food [104]. As of 2,000 in a study carried in Durham, New Hampshire by Chapron et al. [105], adenoviruses together with astroviruses were detected in 51.7 and 48.3% of surface water samples respectively [105]. Adenovirus infections were reported to occur worldwide throughout the year [40], suggesting that there are no seasonal variations in the prevalence of these viruses, thus qualifying these viruses as suitable indicators of human viral pathogens in aquatic environments. Furthermore, PCR-based procedures such as applied real-time PCR that show enough sensitivity to detect not only specific serotypes but also a wide diversity of excreted strains have been described [106]. To this point we can not state exclusively the suitable index for the enteric viruses both in wastewater and drinking waters because there are other proposed indices like Torque teno virus (TTVs) [107], polyomavirus JCPyV [106] which show some degree of suitability as indices. With the increasing popularity of molecular detection methods which are relatively fast and specific compared to the traditional methods such as tissue culture, developing countries may find a solution to the problem of infectious viruses in aquatic environments if such techniques could be incorporated into part of regular monitoring programmes to assess the virus levels in wastewater effluents, and this is a subject of intensive investigation in our group. Microbial Source Tracking (MST) is another promising tool that seeks to predict the source of microbial contamination in the environment, more especially the fecal contamination of aquatic environments [97]. The important aspect of this method is to determine whether the source of fecal contamination is of human or animal origin since viruses are often host-specific [97], and that it may help prevent contamination from its source point. As useful a tool this method may be, it could be negatively influenced by factors like the complexity of the environment under study, the number of sources suspected to be implicated in contamination events, funds available to perform studies, and the technical expertise available to produce and analyze the data, more so in developing countries [97].

We are grateful to the Medical Research Council of South Africa as well as the Govan Mbeki Research and Development Center of the University of Fort Hare for the financial support.

REFERENCES

1. Bosch, A. Human enteric viruses in the water environment: A minireview. Int. Microbiol 1998, 1, 191–196.
2. Carter, MJ. Enterically infecting viruses: Pathogenicity, transmission and significance for food and waterborne infection. J. Appl. Microbiol 2005, 98, 1354–1380.
3. Li, WJ; Xin, WW; Rui, QY; Song, N; Zhang, FG; Ou, YC; Chao, FH. A new and simple method for concentration of enteric viruses from water. J. Virol. Meth 1998, 74, 99–108.
4. Svraka, S; Duizer, E; Vennema, H; de Bruin, E; van der Veer, B; Dorresteijn, B; Koopmans, M. Etiological role of viruses in outbreaks of acute gastroenteritis in The Netherlands from 1994 through 2005. J. Clin. Microbiol 2007, 45, 1389–1394.
5. Cloette, TE; Da Silva, E; Nel, LH. Removal of waterborne human enteric viruses and coliphages with oxidized coal. Curr. Microbiol 1998, 37, 23–27.
6. Karmakar, S; Rathore, AS; Kadri, SM; Dutt, S; Khare, S; Lal, S. Post-earthquake outbreak of rotavirus gastroenteritis in Kashmir (India): An epidemiological analysis. Public Health 2008, 122, 981–989.
7. He, JW; Jiang, S. Quantification of enterococci and human adenoviruses in environmental samples by Real-Time PCR. Appl. Environ. Microbiol 2005, 71, 2250–2255.
8. Weisberg, SS. Rotavirus. Disease-a-Month 2007, 53, 510–514.
9. Dennehy, PH. Rotavirus vaccines-an update. Vaccine 2007, 25, 3137–3141.
10. Madisch, I; Harste, G; Pommer, H; Heim, A. Phylogenetic analysis of the main neutralization and hemagglutination determinants of all human adenovirus prototypes as a basis for molecular classification and taxonomy. J. Virol 2005, 79, 15265–15276.
11. Rahman, M; Hassan, ZM; Zafrul, H; Saiada, F; Banik, S; Faruque, ASG; Delbeke, T; Matthijnssens, J; Van Ranst, M; Azim, T. Sequence analysis and evolution of group B rotaviruses. Virus Res 2007, 125, 219–225.

12. Gómara, MI; Kang, G; Mammen, A; Jana, AK; Abraham, M; Desselberger, U; Brown, D; Gray, J. Characterization of G10P[11] rotaviruses causing acute gastroenteritis in neonates and infants in Vellore, India. J. Clin. Microbiol 2004, 42, 2541–2547.

13. Arista, S; Giammanco, GM; De Grazia, S; Colomba, C; Martella, V. Genetic variability among serotype G4 Italian human rotaviruses. J. Clin. Microbiol 2005, 43, 1420–1425.

14. Burke, B; Desselberger, U. Rotavirus pathogenicity. Virology 1996, 218, 299–305.

15. Lundgren, O; Svensson, L. Pathogenesis of rotavirus diarrhea. Microbes Inf 2001, 3, 1145–1156.

16. Feng, N; Lawton, JA; Gilbert, J; Kuklin, N; Vo, P; Prasad, BVV; Greenberg, HB. Inhibition of rotavirus replication by a non-neutralizing, rotavirus VP6–specific IgA mAb. J. Clin. Invest 2002, 109, 1203–1213.

17. Anderson, EJ; Weber, SG. Rotavirus infection in adults. Lancet Inf. Dis 2004, 4, 91–99.

18. Ramig, RF. Pathogenesis of intestinal and systemic rotavirus infection. J. Virol 2004, 78, 10213–10220.

19. Coulson, BS; Witterick, PD; Tan, Y; Hewish, MJ; Mountford, JN; Harrison, LC; Honeyman, MC. Growth of rotaviruses in primary pancreatic cells. J. Virol 2002, 76, 9537–9544.

20. Steele, AD; Peenze, I; de Beer, MC; Pager, CT; Yeats, J; Potgieter, N; Ramsaroop, U; Page, NA; Mitchell, JO; Geyer, A; Bos, P; Alexander, JJ. Anticipating rotavirus vaccines: epidemiology and surveillance of rotavirus in South Africa. Vaccine 2003, 21, 354–360.

21. Parashar, UD; Gibson, CJ; Bresee, JS; Glass, RI. Rotavirus and severe childhood diarrhea. Emerg. Inf. Dis 2006, 12, 304–306.

22. Ruiz-Palacios, GM; Pérez-Schael, I; Velázquez, FR; Abate, H; Breuer, T; Clemens, SC; Cheuvart, B; Espinoza, F; Gillard, P; Innis, BL; Cervantes, Y; Linhares, AC; López, P; Macías-Parra, M; Ortega-Barría, E; Richardson, V; Rivera-Medina, DM; Rivera, L; Salinas, B; Pavía-Ruz, N; Salmerón, J; Rüttimann, R; Tinoco, JC; Rubio, P; Nuñez, E; Guerrero, ML; Yarzábal, JP; Damaso, S; Tornieporth, N; Sáez-Llorens, X; Vergara, RF; Vesikari, T; Bouckenooghe, A; Clemens, R; De Vos, B; O'Ryan, M; Human Rotavirus Vaccine Study Group. Safety and efficacy of an attenuated vaccine against severe rotavirus gastroenteritis. New Engl. J. Med 2006, 354, 11–22.

23. Nasri, D; Bouslama, L; Omar, S; Saoudin, H; Bourlet, T; Aouni, M; Pozzetto, B; Pillet, S. Typing of human enterovirus by partial sequencing of VP2. J. Clin. Microbiol 2007, 45, 2370–2379.

24. Oberste, MS; Maher, K; Michele, SM; Belliot, G; Uddin, M; Pallansch, MA. Enteroviruses 76, 89, 90 and 91 represent a novel group within the species Human enterovirus A. J. Gen. Virol 2005, 86, 445–451.

25. Oberste, MS; Maher, K; Williams, AJ; Dybdahl-Sissoko, N; Brown, BA; Gookin, MS; Peñaranda, S; Mishrik, N; Uddin, M; Pallansch, MA. Species-specific RT-PCR amplification of human enteroviruses: a tool for rapid species identification of uncharacterized enteroviruses. J. Gen. Virol 2006, 87, 119–128.

26. Palacios, G; Oberst, MS. Enteroviruses as agents of emerging infectious diseases. J. Neur. Virol 2005, 11, 424–433.
27. Williams, CH; Kajander, T; Hyypiä, T; Jackson, T; Sheppard, D; Stanway, G. Integrin vß6 is an RGD-dependent receptor for Coxsackievirus A9. J. Virol 2004, 78, 6967–6973.
28. Bauer, S; Gottesman, G; Sirota, L; Litmanovitz, I; Ashkenazi, S; Levi, I. Severe Coxsackie virus B infection in preterm newborns treated with pleconaril. Eur. J. Ped 2002, 161, 491–493.
29. Peng, T; Li, Y; Yang, Y; Niu, C; Morgan-Capner, P; Archard, LC; Zhang, H. Characterization of Enterovirus Isolates from patients with muscle disease in a Selenium-deficient area in China. J. Clin. Microbiol 2000, 38, 3538–3543.
30. Hyypia, T; Hovi, T; Knowles, NJ; Stanway, G. Classification of enteroviruses based on molecular and biological properties. J. Gen. Virol 1997, 78, 1–11.
31. Daley, JK; Gechman, LA; Skipworth, J; Rall, GF. Poliovirus replication and spread in primary neuron cultures. Virology 2005, 340, 10–20. [Google Scholar]
32. Racaniello, VR. One hundred years of poliovirus pathogenesis. Virology 2006, 344, 9–16.
33. Vellinga, J; Van der Heijdt, S; Hoeben, RC. The adenovirus capsid: major progress in minor proteins. J. Gen. Virol 2005, 86, 1581–1588.
34. Kajon, AE; Moseley, JM; Metzgar, D; Huong, HS; Wadleigh, A; Ryan, MAK; Russell, KL. Molecular epidemiology of adenovirus type 4 infections in US military recruits in the Postvaccination Era (1997–2003). J. Inf. Dis 2007, 196, 67–75.
35. Goosney, DL; Nemerow, GR. Adenovirus infection: taking the back roads to viral entry. Current Biol 2003, 13, 99–100.
36. Zubieta, C; Schoehn, G; Chroboczek, J; Cusack, S. The structure of the human adenovirus 2 Penton. Mol.Cell 2005, 17, 319–320.
37. Coyne, CB; Bergelson, JM. CAR: a virus receptor within the tight junction. Adv. Drug Deli. Revs 2005, 57, 869–882.
38. Walters, RW; Freimuth, P; Moninger, TO; Ganske, I; Zabner, J; Welsh, MJ. Adenovirus fiber disrupts CAR-mediated intercellular adhesion allowing virus escape. Cell 2002, 110, 789–799.
39. Echavarria, M; Forman, M; van Tol, MJD; Vossen, JM; Charache, P; Kroes, ACM. Prediction of severe disseminated adenovirus infection by serum PCR. Lancet 2001, 358, 384–385.
40. Flomenberg, P. Adenovirus infections. Medicine 2005, 33, 128–130.
41. Chakravarty, S; Hutson, AM; Estes, MK; Prasad, BVV. Evolutionary trace residues in noroviruses: importance in receptor binding, antigenicity, virion assembly, and strain diversity. J. Virol 2005, 79, 554–568.
42. Butt, AA; Aldridge, KE; Sanders, CV. Infections related to the ingestion of seafood Part I: viral and bacterial infections. Lancet Inf. Dis 2004, 4, 201–212.
43. Prasad, BVV; Hardy, ME; Dokland, T; Bella, J; Rossmann, MG; Estes, MK. X-ray crystallographic structure of the norwalk virus capsid. Science 1999, 286, 287–290.
44. Tan, M; Meller, J; Jiang, X. C-Terminal Arginine cluster is essential for receptor binding of Norovirus Capsid Protein. J. Virol 2006, 80, 7322–7331.

45. Hutson, AM; Atmar, RL; Estes, MK. Norovirus disease: changing epidemiology and host susceptibility factors. Trends Mic 2004, 12, 279–287.

46. Maguire, AJ; Green, J; Brown, DWG; Desselberger, U; Gray, JJ. Molecular epidemiology of outbreaks of Gastroenteritis associated with small round-structured viruses in east Anglia, United Kingdom, during the 1996–1997 season. J. Clin. Microbiol 1999, 37, 81–89.

47. Ike, AC; Brockmann, SO; Hartelt, K; Marschang, RE; Contzen, M; Oehme, RM. Molecular epidemiology of norovirus in outbreaks of Gastroenteritis in southwest Germany from 2001 to 2004. J. Clin. Microbiol 2006, 44, 1262–1267.

48. Argaw, N. Chapter 6: Wastewater Sources and Treatment. In Renewable Energy in Water and Wastewater Treatment Applications; National Renewable Energy Laboratory, US Department of Energy Laboratory: Golden, CO, USA, 2004; pp. 38–46.

49. DeBusk, WF. Wastewater treatment wetlands: contaminant removal processes; University of Florida, IFAS Extension: Gainesville, FL, USA, 1999; pp. 1–5.

50. Templeton, MR; Andrews, RC; Hofmann, R. Particle-associated viruses in water: impacts on disinfection processes. Critical Revs. Environ. Sci. Tech 2008, 38, 137–164.

51. Templeton, M; Hofmann, R; Andrews, RC. Ultraviolet disinfection of particle-associated viruses. In Chemical Water and Wastewater Treatment; Hahn, H, Hoffman, E, Odegaard, H, Eds.; IWA Publishing: Padstow, Cornwall, UK, 2004; pp. 109–116.

52. Ueda, T; Horan, NJ. Fate of indigenous bacteriophage in a membrane bioreactor. Water Res 2004, 34, 2151–2159.

53. Mezzanotte, V; Antonelli, M; Citterio, S; Nurizzo, C. Wastewater disinfection alternatives: Chlorine, Ozone, Peracetic Acid, and UV Light. Water Environ. Res 2007, 79, 2373–2379.

54. Formiga-Cruz, M; Tofiño-Quesada, G; Bofill-Mas, S; Lees, DN; Henshilwood, K; Allard, AK; Conden-Hansson, AC; Hernroth, BE; Vantarakis, A; Tsibouxi, A; Papapetropoulou, M; Furones, MD; Girones, R. Distribution of human virus contamination in shellfish from different growing areas in Greece, Spain, Sweden, and the United Kingdom. Appl. Environ. Microbiol 2002, 68, 5990–5998.

55. Fong, TT; Lipp, EK. Enteric viruses of humans and animals in aquatic environments: health risks, detection, and potential water quality assessment tools. Microbiol. Mol. Biol. Revs 2005, 69, 357–371.

56. Maunula, L; Miettinen, IT; von Bonsdorff, CH. Norovirus outbreaks from drinking water. Emerg. Inf. Dis 2007, 11, 1716–1721.

57. Igbinosa, EO; Obi, LC; Okoh, AI. Occurrence of potentially pathogenic vibrios in the final effluents of a wastewater treatment facility in a rural community of the eastern Cape Province of South Africa. Res. Microbiol 2009, 160, 531–537.

58. Odjadjare, EEO; Okoh, AI. Prevalence and distribution of Listeria pathogens in the final effluents of a rural wastewater treatment facility in the eastern Cape Province of South Africa. World J. Microbiol. & Biotech 2010, 26, 297–307.

59. Okoh, AI; Odjadjare, EE; Igbinosa, EO; Osode, AN. Wastewater treatment plants as a source of microbial pathogens in the receiving watershed. Afr. J. Biotech 2007, 6, 2932–2944.

60. Igbinosa, EO; Okoh, AI. Impact of discharge wastewater effluents on the physic-chemical qualities of a receiving watershed in a typical rural community. Int. J. Environ. Sci. Tech 2009, 6, 175–182.

61. Ehlers, MM; Grabow, WOK; Pavlov, DN. Detection of enteroviruses in un-treated and treated drinking water supplies in South Africa. Water Res 2005, 39, 2253–2258.

62. Vivier, JC; Ehlers, MM; Grabow, WOK. Detection of enteroviruses in treated drinking water. Water Res 2004, 38, 2699–2705.

63. Zhang, K; Farahbakhsh, K. Removal of native coliphages and coliform bacteria from municipal wastewater by various wastewater treatment processes: Implica-tions to water reuse. Water Res 2007, 41, 2816–2824.

64. Arraj, A; Bohatier, J; Laveran, H; Traore, O. Comparison of bacteriophage and enteric virus removal in pilot scale activated sludge plants. J. Appl. Microbiol 2005, 98, 516–524.

65. Charles, KJ; Souter, FC; Baker, DL; Davies, CM; Schijven, JF; Roser, DJ; Deere, DA; Priscott, PK; Ashbolt, NJ. Fate and transport of viruses during sewage treat-ment in a mound system. Water Res 2008, 42, 3047–3056.

66. Gersberg, RM; Lyon, SR; Brenner, R; Elkins, BV. Performance of a clay-alum flocculation (CCBA) process for virus removal from municipal wastewater. Water Res 1988, 22, 1449–1454.

67. Zhu, B; Clifford, DA; Chellam, S. Virus removal by iron coagulation–microfiltra-tion. Water Res 2005, 39, 5153–5161.

68. Schijven, JF; Hassanizadeh, SM. Removal of viruses by soil passage: overview of modeling, processes and parameters. Critical Revs. Environ. Sci. & Tech 2000, 30, 49–127.

69. Olson, MR; Axler, RP; Hicks, RE; Henneck, JR; McCarthy, BJ. Seasonal virus removal by alternative onsite wastewater treatment systems. J. Water Health 2005, 3, 139–155.

70. Cho, M; Chung, H; Yoon, J. Disinfection of water containing natural organic mat-ter by using Ozone-initiated radical reactions. Appl. Environ. Microbiol 2003, 69, 2284–2291.

71. Kurosaki, Y; Abe, H; Morioka, H; Hirayama, J; Ikebuchi, K; Kamo, N; Nikaido, O; Azuma, H; Ikeda, H. Pyrimidine Dimer formation and Oxidative damage in M13 bacteriophage inactivation by Ultraviolet C irradiation. Photochem. Photo-biol 2007, 78, 349–354.

72. Gehr, R; Wagner, M; Veerasubramanian, P; Payment, P. Disinfection efficiency of peracetic acid, UV and ozone after enhanced primary treatment of municipal wastewater. Water Res 2003, 37, 4573–4586.

73. Templeton, MR; Andrews, RC; Hofmann, R. Removal of particle-associated bac-teriophages by dual-media filtration at different filter cycle stages and impacts on subsequent UV disinfection. Water Res 2007, 41, 2393–2406.

74. Tree, JA; Adams, MR; Lees, DN. Chlorination of indicator bacteria and viruses in primary sewage effluent. Appl. Environ. Microbiol 2003, 69, 2038–2043.

75. Armon, R; Cabelli, VJ; Duncanson, R. Survival of F-RNA Coliphages and three bacterial indicators during wastewater chlorination and transport in estuarine wa-ters. Estuaries Coasts 2007, 30, 1088–1094.

76. Meng, QS; Gerba, CP. Comparative inactivation of enteric adenoviruses, polio-virus and coliphages by ultraviolet irradiation. Water Res 1996, 30, 2665–2668.

77. Thurston-Enriquez, JA; Haas, CN; Jacangelo, J; Gerba, CP. Inactivation of enteric adenovirus and feline Calicivirus by chlorine dioxide. Appl. Environ. Microbiol 2005, 71, 3100–3105.

78. Griffin, DW; Donaldson, KA; Paul, JH; Rose, JB. Pathogenic human viruses in coastal waters. Clin. Microbiol. Revs 2003, 16, 129–143.

79. Goyal, SM; Adams, WN; O'Malley, ML; Lear, DW. Human pathogenic viruses at sewage sludge disposal sites in the Middle Atlantic region. Appl. Environ. Micro-biol 1984, 48, 758–763.

80. Borchardt, MA; Bertz, PD; Spencer, SK; Battigelli, DA. Incidence of enteric vi-ruses in groundwater from household wells in Wisconsin. Appl. Environ. Micro-biol 2003, 69, 1172–1180.

81. van Heerden, J; Ehlers, MM; Grabow, WOK. Detection and risk assessment of adenoviruses in swimming pool water. J. Appl. Microbiol 2005, 99, 1256–1264.

82. Fout, GS; Martinson, BC; Moyer, MWN; Dahling, DR. A multiplex reverse tran-scription-PCR method for detection of human enteric viruses in groundwater. Appl. Environ. Microbiol 2003, 69, 3158–3164.

83. Xagoraraki, I; Kuo, DHW; Wong, K; Wong, M; Rose, JB. Occurrence of human adenoviruses at two recreational beaches of the great lakes. Appl. Environ. Micro-biol 2007, 73, 7874–7881.

84. Mocé-Llivina, L; Lucena, F; Jofre, J. Enteroviruses and bacteriophages in bathing Waters. Appl. Environ. Mic 2005, 71, 6838–6844.

85. Kukkula, M; Maunula, L; Silvennoinen, E; von Bonsdorff, CH. Outbreak of viral gastroenteritis due to drinking water contaminated by Norwalk-like viruses. J. Inf. Dis 1999, 180, 1771–1776.

86. Le Guyader, FS; Mittelholzer, C; Haugarreau, L; Hedlund, KO; Alsterlund, R; Pommepuy, M; Svensson, L. Detection of noroviruses in raspberries associated with a gastroenteritis outbreak. Int. J. Food Microbiol 2004, 97, 179–186.

87. Grotto, I; Huerta, M; Balicer, RD; Halperin, T; Cohen, D; Orr, N; Gdalevich, M. An outbreak of Norovirus gastroenteritis on an Israeli military base. Infection 2004, 32, 339–343.

88. Harris, LJ; Farber, JN; Beuchat, LR; Parish, ME; Suslow, TV; Garrett, EH; Busta, FF. Outbreaks associated with fresh produce: incidence, growth, and survival of pathogens in fresh and fresh-cut produce. Comp. Revs. Food Sci. Food Saf 2006, 2, 78–141.

89. Richards, GP. Food-borne pathogens. Enteric virus contamination of foods through industrial practices: a primer on intervention strategies. J. Ind. Microbiol. Biotech 2001, 27, 117–125.

90. Chironna, M; Lopalco, P; Prato, R; Germinario, C; Barbuti, S; Quarto, M. Out-break of infection with Hepatitis A Virus (HAV) associated with a foodhandler and confirmed by sequence analysis reveals a new HAV Genotype IB variant. J. Clin. Microbiol 2004, 42, 2825–2828.

91. Martinez, A; Dominguez, A; Torner, N; Ruiz, L; Camps, N; Barrabeig, I; Arias, C; Alvarez, J; Godoy, P; Balaña, PJ; Pumares, A; Bartolome, R; Ferrer, D; Perez,

U; Pinto, R; Buesa, J. The Catalan viral Gastroenteritis study group: epidemiology of foodborne norovirus outbreaks in Catalonia, Spain. BMC Inf. Dis 2008, 8, 47.

92. Le Guyader, FS; Bon, F; DeMedici, D; Parnaudeau, S; Bertone, A; Crudeli, S; Doyle, A; Zidane, M; Suffredini, E; Kohli, E; Maddalo, F; Monini, M; Gallay, A; Pommepuy, M; Pothier, P; Ruggeri, FM. Detection of multiple noroviruses associated with an international gastroenteritis outbreak linked to oyster consumption. J. Clin. Microbiol 2006, 44, 3878–3882.

93. Rehnstam-Holm, AS; Hernroth, B. Shellfish and public health: A Swedish perspective. AMBIO: J. Human Environ 2005, 34, 139–144.

94. Sincero, TCM; Levin, DB; Simões, CMO; Barardi, CRM. Detection of hepatitis A virus (HAV) in oysters (Crassostrea gigas). Water Res 2006, 40, 895–902.

95. Huppatz, C; Munnoch, SA; Worgan, T; Merritt, TD; Dalton, C; Kelly, PM; Durrheim, DN. A norovirus outbreak associated with consumption of oysters: implications for quality assurance systems. CDI 2008, 32, 88–91.

96. Karamoko, Y; Ibenyassine, K; Mhand, RA; Idaoma, M; Ennaji, MM. Assessment of enterovirus contamination in mussel samples from Morocco. World J. Microbiol. Biotech 2006, 22, 105–108.

97. Santo Domingo, JW; Edge, TA. Identification of primary sources of faecal pollution. In Safe Management of Shellfish and Harvest Waters; Rees, G, Pond, K, Kay, D, Bartram, J, Santo Domingo, J, Eds.; IWA Publishing: London, UK, 2010; pp. 51–90.

98. Halliday, ML; Kang, LY; Zhou, TK; Hu, MD; Pan, QC; Fu, TY; Huang, YS; Hu, SL. An epidemic of hepatitis a attributable to the ingestion of raw clams in Shanghai, China. J. Infect. Dis 1991, 164, 852–859.

99. Tang, YW; Wang, JX; Xu, ZY; Guo, YF; Qian, WH; Xu, JX. A serological confirmed, case-control study, of a large outbreak of hepatitis A in China, associated with consumption of clams. Epidemiol. Inf 1991, 107, 651–657.

100. Potasman, I; Paz, A; Odeh, M. Infectious outbreak associated with bivalve shellfish consumption: a worldwide perspective. Clin. Inf. Dis 2002, 35, 921–928.

101. Sair, AI; D'Souza, DH; Jaykus, LA. Human enteric viruses as causes of foodborne diseases. Comp. Revs. Food Sci. Food Saf 2002, 1, 73–89.

102. von Sperling, M; Chernicharo, CAL. Urban wastewater treatment technologies and the implementation of discharge standards in developing countries. Urban Water 2002, 4, 105–114.

103. Girones, R. Tracking viruses that contaminate environments using PCR to track stable viruses provides an effective means for monitoring water quality for environmental contaminants. Microbe 2006, 1, 19–25.

104. Dongdem, JT; Soyiri, I; Ocloo, A. Public health significance of viral contamination of drinking water. Afr. J. Microbiol. Res 2009, 3, 856–861.

105. Chapron, CD; Ballester, NA; Fontaine, JH; Frades, CN; Margolin, AB. Detection of astroviruses, enteroviruses and adenovirus types 40 and 41 in surface waters collected and evaluated by the information collection rule and an integrated cell culture-nested PCR procedure. Appl. Environ. Microbiol 2000, 66, 2520–2525.

106. Bofill-Mas, S; Albinana-Gimenez, N; Clemente-Casares, P; Hundesa, A; Rodriguez-Manzano, J; Allard, A; Calvo, M; Girones, R. Quantification and stability

of human adenoviruses and Polyomavirus JCPyV in wastewater matrices. Appl. Environ. Microbiol 2006, 72, 7894–7896.

107. Griffin, JS; Plummer, JD; Long, SC. Torque teno virus: an improved indicator for viral pathogens in drinking waters. Virol. J 2008, 5, 1–6.

CHAPTER 2

Prevalence and Characterisation of Non-Cholerae *Vibrio* spp. in Final Effluents of Wastewater Treatment Facilities in Two Districts of the Eastern Cape Province of South Africa: Implications for Public Health

ANTHONY I. OKOH, TIMOTHY SIBANDA, VUYOKAZI NONGOGO, MARTINS ADEFISOYE, OSUOLALE O. OLAYEMI, AND NOLONWABO NONTONGANA

2.1 INTRODUCTION

Municipal wastewater, even if treated, may still contain a wide range of pathogens that are excreted by diseased humans (Arceivala 1997; Wen et al. 2009), contributing to increased densities of pathogens in the receiving water bodies. Surface water bodies are major sources of potable water whose unabated contamination has led to many water-related disease outbreaks in the past (Mishra et al. 2004; Nair et al. 2004; Obi et al. 2004). Although wastewater treatment technologies can, with optimised performance, reduce bacterial and viral pathogens by approximately 90% (Asano and Levine 1998; Jiménez et al. 2004), it is not possible for the microbial quality of the effluents to match the microbial quality of the water in the receiving water bodies. Discharge of effluents will, therefore, despite the level of treatment, potentially alter the microbial content of the receiving water bodies (Drury et al. 2013). Previous studies show that some *Vibrio* species survive the activated sludge-based wastewater treatment process as free cell and as plankton-associated entities (Igbinosa et al. 2009, 2011), suggesting that the provision of wastewater treatment facilities does not, in itself, ensure satisfactory effluent water quality. The most common clinical presentation of *Vibrio* infection is self-limited gastroenteritis, but wound infections and primary septicaemia may also occur (Levine and Griffin 1993). Healthy carriers of *Vibrio cholerae* excrete *Vibrios* intermittently, with chronic convalescent carriers shedding *Vibrios* intermittently for periods of 4 to 15 months (Nevondo and Cloete 2001). Survival of *Vibrios* in the aquatic environment relates sharply to various chemical, biological and physical characteristics of the aquatic milieu, with *V. cholerae* known to remain viable in surface waters for periods ranging from 1 h to 13 days, while faecal contamination from victims of epidemics and healthy carriers may continue to reinforce their concentrations in water (Nevondo and Cloete 2001). As a result, cholera and cholera-like infections continue to be a substantial health burden in developing countries, especially in Africa and Asia, compromising the primary health of vulnerable members of society (Bourne and Coetzee 1996; Pegram et al. 1998; Mackintosh and Colvin 2003). *Vibrio* species have been incriminated in cases of diarrhoea, accounting for a substantial degree of morbidity and mortality in different age groups worldwide (Obi et al. 2004). The most notable of *Vibrio* pathogens are *V. cholerae*, *Vibrio parahaemolyticus*, *Vibrio vulnificus* and *Vibrio fluvialis* (CDC Centers for Disease Control and Prevention 1999; Finkelstein et al. 2002; Kothary et al. 2003; Chakraborty et al. 2006) which are mainly transmitted via water and food. They all cause diarrhoea, but in entirely different ways; *V. vulnificus* and *V. parahaemolyticus* are invasive organisms, affecting primarily the colon, while *V. cholerae* is non-invasive, affecting the small intestine through secretion of an enterotoxin (Todar 2005), and is the etiologic agent of cholera. The clinical symptoms of *V. fluvialis* gastroenteritis are similar to cholera with the additional manifestation of bloody stools which is suggestive of an invasive pathogen (Oliver and Kaper 2001). Other *Vibrios* like *Vibrio alginolyticus, Vibrio cincinnatiensis, Vibrio furnisii, Vibrio harveyi, Vibrio metschnikovii* and *Vibrio mimicus* have occasionally been reported as causes of human infections (Farmer and Hickman-Brenner 1992; Abbott and Janda 1994; Carnahan et al. 1994). However, of all the *Vibrio* species which have been associated with illness in humans, the most important are *V. cholerae* subgroups O1 and O139, the causative agents of epidemic cholera (Heymann 2008). Heidelberg et al. (2002) have reported large numbers of *Vibrios*, about $4.3 \times 10^6/mm^2$, attached to the external surface of plankton (zooplankton and phytoplankton), pointing to a close association between *Vibrios* and planktons. This association has also been observed in municipal wastewaters (Ahmadi et al. 2005; Chindah et al. 2007; Mukhopadhyay et al. 2007). *V. fluvialis*, in particular, has been identified as an important cause of cholera-like bloody diarrhoea and primary septicaemia in immunocompromised individuals, especially in underdeveloped countries with poor sanitation (Igbinosa and Okoh 2010). This organism has been isolated from treated wastewater effluents in South Africa (Igbinosa et al. 2009), and there are reports linking it to food poisoning (Kobayashi and Ohnaka 1989), especially due to consumption of raw shellfish (Levine and Griffin 1993). While adequate and timely rehydration therapy remains the gold-standard treatment for cholera and cholera-like bloody diarrhoea (Heymann 2008), antimicrobials are also prescribed for the management of severe cases in order to shorten the duration of illness and reduce the volume of rehydration solution required. However, some *Vibrio* strains are resistant to a number of antimicrobials including tetracycline, co-trimoxazole, trimethoprim and sulfamethoxazole. This resistance to antimicrobials, in addition to other properties such as virulence factors and ability to cause epidemics, makes *Vibrios* pathogens of public health concern. Knowledge of the prevalence and antimicrobial resistance profile of local strains is, therefore, important for the management of complicated cases in the case of an epidemic. At times, a cholera outbreak is reported without any clear linkage of the index case to neighbouring countries or travel to affected areas. This usually leaves health authorities asking themselves where the cholera causing bacteria could have come from. We hypothesise that such outbreaks could be related to either persistence of organisms in free-living, altered or adapted forms capable of reverting to a pathogenic variety or to continuous year-round transmission by sub-clinical cases or a combination of both. Routine analysis of the microbial quality of treated wastewater effluents is therefore warranted in order to maintain the microbial load of receiving water bodies within acceptable limits for both human use and lotic ecosystems survival. There is, however, paucity of information on the molecular epidemiology of *Vibrios* in the aquatic milieu of the Eastern Cape Province (ECP) of South Africa. Compounding the challenge is the production of poor-quality effluents by wastewater treatment plants in the ECP which is acknowledged as mostly non-urban, poor and without adequate infrastructure (Mohale 2003; BLACKSASH 2010). Also, the documentation of final effluent compliance of the wastewater treatment plants to set guidelines with respect to bacteriological quality remains poor in the province. This study was, therefore, aimed at assessing the prevalence of non-cholerae *Vibrios* in the final effluents of 14 wastewater treatment facilities in the ECP of South Africa, characterising them into species and determining their antibiogram properties and evaluating the public health implications of the findings. While there are at least 12 pathogenic *Vibrio* species recognised to cause human illness (Janda et al. 1988), this work was based on the prevalence and characterisation of *V. vulnificus*, *V. fluvialis* and *V. parahaemolyticus*.

2.2 METHODOLOGY

2.2.1 DESCRIPTION OF STUDY AREA

Fourteen (14) wastewater treatment plants (WWTP) were selected in Amathole and Chris Hani district municipalities of the ECP in South Africa (Fig. 1). The choice of WWTP was influenced by the need to ensure that plants were not located more than 3-h drive from the University of Fort Hare in Alice, such that samples could be taken to the laboratory for analysis within 6 h of collection. The ECP borders the provinces of the Western Cape, Northern Cape, Free State and KwaZulu-Natal, as well as Lesotho in the north. The province is mostly rural with a high percentage of people living in poverty (67.4 %) and a very low Human Development Index (HDI) of 0.52 (BLACKSASH 2010; ECSECC 2011). It is the second largest province in South Africa and mainly comprised of rural

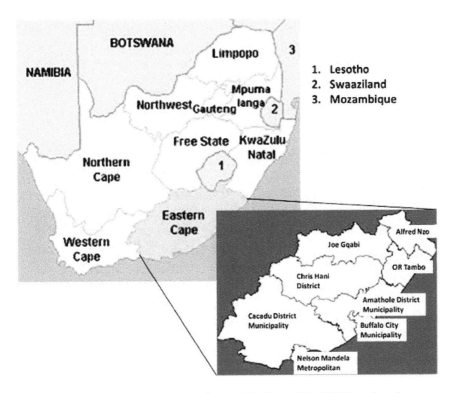

Figure 1. A map showing the seven district municipalities of the ECP. Sampling sites were selected in Chris Hani and Amathole district municipalities (http://www.worldlicenseplat es.com/world/AF_ZAEC.html).

settlements with little or no adequate sanitary facilities, with about 36 % of the population directly reliant on surface water sources for domestic use (ECSECC 2011). The ECP is divided into seven district municipalities, namely, Alfred Nzo, Amathole, Chris Hani, Joe Gqabi, O.R. Tambo, Cacadu and the Nelson Mandela Metropolitan Municipality. The Amathole District Municipality includes the Buffalo City Municipality. Due to the confidential nature of this work, the sampling sites were designated as WWTP1 to WWTP14, and geographical coordinates could equally not be given. All selected WWTPs discharge their treated effluents directly into rivers. Samples were collected with permission from the Amathole and Chris Hani district municipalities.

2.2.2 SAMPLE COLLECTION AND ISOLATION OF PRESUMPTIVE VIBRIO ORGANISMS

Wastewater final effluent samples were collected once a month for a period of 12 months starting in August 2012 to July 2013. Samples were collected in sterile 2-l polypropylene bottles containing 0.1 % of a 3 % (w/v) solution of sodium thiosulphate for sample dechlorination and taken to the Applied and Environmental Microbiology Research Group (AEMREG) laboratory at the University of Fort Hare in Alice, South Africa, in cooler boxes containing ice, for analysis within 6 h of collection. Samples were serially diluted and concentrated on nitrocellulose membrane filters (0.45-μm pore size, Millipore) by passing 100 ml of each dilution through the filter using the membrane filtration technique as recommended by Standard (2005). The filters were then placed onto agar plates containing thiosulphate citrate bile salts sucrose agar (TCBS agar). For the purposes of quality control, the spread plate technique was also employed where known (100 μl) volumes of effluent samples were spread on TCBS agar as previously described by Igbinosa et al. (2011). Green and yellow colonies were identified and enumerated as presumptive *Vibrio* isolates. Counts were converted to Log10 values and clustered into seasons where spring composed of August 2012–October 2012, summer (November 2012–January 2013), autumn (February 2013, March 2013 and April 2013) and winter (May 2013–July 2013). Presumptive *Vibrio* colonies were then isolated, purified and subjected to Gram staining and oxidase test. Gram-negative, oxidase positive isolates were selected and preserved in 20 % glycerol at −80 °C until further analysis.

2.2.3 MOLECULAR CONFIRMATION OF VIBRIO SPECIES

To extract DNA, single colonies of 18–24 h old presumptive *Vibrio* cultures grown on nutrient agar plates at 37 °C were picked, suspended in

200 µl of sterile distilled water and the cells lysed using an AccuBlock (Digital dry bath, Labnet) for 15 min at 100 °C as described by Maugeri et al. (2006). The cell debris was removed by centrifugation at 11,000×g for 2 min using a MiniSpin microcentrifuge (Lasec, RSA). Five microlitre (5 µl) aliquots of the cell lysates were used as template DNA in polymerase chain reaction (PCR) assays immediately after extraction to determine the molecular identity of the isolates. To confirm if the isolates belonged to the genus *Vibrio*, the primer set V16S-700F (CGG TGA AAT GCG TAG AGA T) and V16S-1325R (TTA CTA GCG ATT CCG AGT TC) targeting the 16S ribosomal RNA (rRNA) gene (663 bp) was used in PCR assays as described by Kwok et al. (2002).

Vibrio species identification was done using species-specific primers targeting specific sequences within the 16S rRNA as described by Kim et al. (1999). For *V. parahaemolyticus*, the primer set Vp. flaE-79F (GCA GCT GAT CAA AAC GTT GAG T) and Vp. flaE-934R (ATT ATC GAT CGT GCC ACT CAC) targeting the 897 bp flaE gene was used as described by Tarr et al. (2007). *V. vulnificus* was identified using the primer set Vv. hsp-326F (GTC TTA AAG CGG TTG CTG C) and Vv. hsp-697R (CGC TTC AAG TGC TGG TAG AAG) targeting the hsp60 gene (410 bp) as described by Wong and Chow (2002), while the primer set Vf toxR-F (GAC CAG GGC TTT GAG GTG GAC GAC) and Vf toxR-R (AGG ATA CGG CAC TTG AGT AAG ACTC) was used to identify *V. fluvialis* targeting the 217 bp toxR gene as previously described (Osorio and Klose 2000; Chakraborty et al. 2006). In all, 300 confirmed *Vibrio* isolates were randomly selected from a pool of 668 confirmed *Vibrio* isolates taken from across all sampling sites and characterised into these three pathotypes using PCR. *V. parahaemolyticus* DSM 11058, *V. fluvialis* DSM 19283 and *V. vulnificus* DSM 11507 were used as positive control strains.

2.2.4 ANTIBIOGRAM CHARACTERISATION

All *Vibrio* isolates that were positively identified to belong to any of the three pathotypes were subjected to antimicrobial susceptibility testing using the following antibiotics: imipenem, nalidixic acid, erythromycin, sulfamethoxazole, cefuroxime, penicillin G, chloramphenicol, polymixin

B, trimethoprim-sulfamethoxazole, tetracycline, gentamicin, meropenem and trimethoprim. Ciprofloxacin and doxycycline have been the antibiotics of choice for adults (except pregnant women) (Steinberg et al. 2001), while erythromycin and trimethoprim-sulfamethoxazole have been recommended for children and pregnant women. These antimicrobial agents are, however, no longer recommended as first-line therapy because of increasing global antimicrobial resistance (Gilbert et al. 1999), and whenever possible, treatment protocols should be based on local antibiogram data (Centers for Disease Control and Prevention 2013).

2.2.5 STATISTICAL ANALYSIS

An independent samples t test (IBM SPSS version 20) was used to compare the mean presumptive *Vibrio* counts of all the WWTPs and also the mean presumptive *Vibrio* counts obtained in each of the seasons. Differences were deemed significant at $P < 0.05$.

2.3 RESULTS AND DISCUSSION

Presumptive *Vibrio* organisms were isolated in all seasons and at all sampling sites with the exception of WWTP10 where presumptive *Vibrio* was not detected in summer and spring. Gastrointestinal pathogenic microorganisms do not occur as a natural part of the normal intestinal microbiota (Gerritsen et al. 2011). Their presence in wastewater, therefore, could be dependent on the number of infected people in the population contributing to the wastewater flow. The presumptive *Vibrio* counts for all sampling sites were expressed in Log10 values and are presented in Figs. 2 and 3. The error bars on these figures represent the standard deviations since each of the readings is an average of the counts of 3 months constituting each season. Significantly higher presumptive *Vibrio* counts were obtained in samples from WWTP2 ($P < 0.05$), while samples from WWTP10 had significantly lower counts compared to the rest of the WWTPs. While other studies have reported a reduction in environmental *Vibrio* densities during winter as compared to other seasons (DePola et al. 2003; de Souza

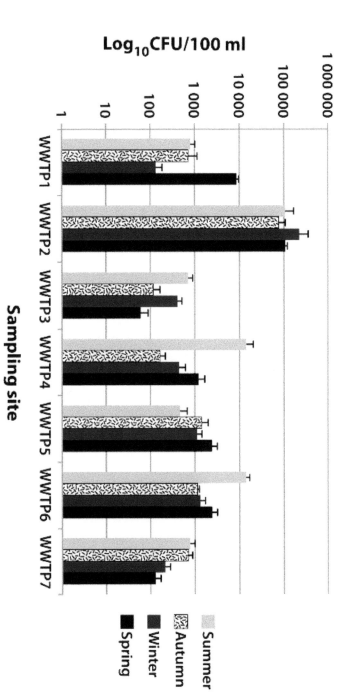

Figure 2. Seasonal presumptive *Vibrio* counts for WWTP1-7.

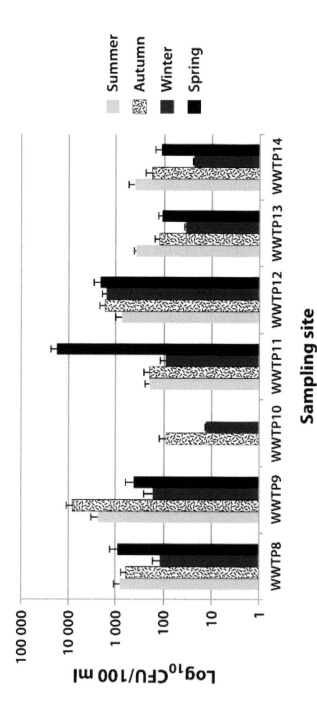

Figure 3. Seasonal presumptive Vibrio counts for WWTP8-14

Costa Sobrinho et al. 2010), the trend was different in our case as statistical analysis showed that there was no significant difference in *Vibrio* densities obtained in different seasons.

South Africa has been plagued by outbreaks of *Vibrio*-related waterborne infections that are suspected to be linked to inefficiently treated effluents discharge from wastewater treatment facilities (Igbinosa et al. 2011).

When a randomised sample of 1,000 presumptive *Vibrio* isolates were subjected to PCR confirmation for *Vibrio* organisms, 668 isolates tested positive, giving an overall *Vibrio* prevalence rate of 66.8 %. As at the time when this study was carried out, there was no known outbreak of cholera or cholera-like diarrhoea in the ECP and, specifically, in the two districts serviced by the selected 14 WWTPs and yet *Vibrio* was isolated all the same. *Vibrio* species are not normal biota for human beings, as some *E. coli* are, neither are they normal biota for fresh water environments, and positive results from this work strongly point to either unreported sporadic incidents of infection within the communities or the existence of healthy *Vibrio* carriers intermittently shedding *Vibrios* into the environment as suggested by Nevondo and Cloete (2001). Similar findings were obtained by Jackson S. Beney C (2000) who managed to detect potentially virulent *V. cholerae* in freshwater environments despite the absence of clinical cases in the host population for some time. Harris et al. (2012) also stated that some patients can even be infected with *V. cholerae* O1 or O139 and yet show no symptoms but then tend to shed the organism into the environment, even for only a few days, explaining why *Vibrios* can be isolated in wastewater effluents in a non-*Vibrio* and/or non-cholera epidemic area. Once these *Vibrios* get into environmental water, they convert to conditionally viable environmental cells within 24 h (Faruque et al. 2005; Nelson et al. 2008). Such *Vibrios* are infectious on reintroduction into a human body, but the infectious dose in this form is not known (Harris et al. 2012). This becomes a major public health time bomb in underdeveloped areas like the ECP where, as of 2011, about 36 % of the population still got their drinking water directly from rivers and streams (ECSECC 2011). An excerpt from Water Supply, Sanitation and Hygiene (WASH 2010) summed up the challenge as follows,

"Many of South Africa's municipal wastewater treatment plants (WWTP) are not performing to acceptable water quality standards .... A lack of good-quality drinking water leads to health problems, which is serious, given the fact that many poor citizens source water directly from the rivers, where not only municipalities .... Since South Africa does not have large rivers, the discharged effluents concentrate into small watercourses"

If, on the average, every litre of effluent contains about 1,000 presumptive *Vibrio* organisms, as was the case in this study, and the lowest effluent volume produced by a treatment plant is 0.63 ml/day (WWTP7), there will be a daily addition of about 6.3×10^8 presumptive *Vibrio* cells into the environment. The resultant health risk to those who drink untreated water directly from the receiving watercourses can be exacerbated if the receiving watercourses are small as this will minimise the dilution capacity and result in concentration of potentially virulent organisms, increasing the risk of illness in the case of raw water ingestion. While there is no available data to show the proportion of effluent in the receiving rivers' annual flow volumes in the ECP or in South Africa in general, the annual flow of the Chicago Area Waterway System (which includes all segments of the Chicago River as well as the North Shore Channel) comprises more than 70 % of treated municipal wastewater effluent (Illinois Department of Natural Resources 2011). While the WWTPs in the ECP may not be as numerous and big as to contribute to as high a percentage of the river flow volume, chances are that they may not be as efficient in pathogen removal, taking into account the excerpt from WASH (2010). Other factors that may increase or upset the risk include the infectious dose of the organisms, the presence or absence of virulent factors in the said *Vibrios* and whether or not people filter their water at home before they drink it. Besides direct consumption of untreated surface water, *Vibrios* concentrate in the gut of filter-feeders such as oysters, clams and mussels, where they multiply (Iwamoto et al. 2010). While thorough cooking will destroy these organisms, oysters are often eaten raw and are the most common food associated with *V. parahaemolyticus* infection in the USA (Hlady 1997). Our findings, though not complemented by assessments for virulence and antibiotic resistance genes, indicate that the potential for disease in the community and call for pro-active rather than reactive measures if public health is to be preserved and unnecessary loss of life avoided.

When 300 of the 668 confirmed *Vibrio* isolates were further screened into species, 11.6 % (35) were confirmed to be *V. parahaemolyticus*, 28.6 % (86) were confirmed to be *V. fluvialis* and 28 % (84) were confirmed to be *V. vulnificus*, while 31.8 % (95) belonged to other *Vibrio* spp. not assayed for in this study. When these confirmed *Vibrio* pathotypes were subjected to antibiogram profiling, *V. fluvialis* showed ≥50 % resistance to 8 of the 13 antibiotics used while showing ≥50 % susceptibility to only 4 antibiotics used (Fig. 4).

Similar results have been reported by Okoh and Igbinosa (2010) whose work showed 100, 92, 90, 70 and 80 % resistances by *V. fluvialis* isolates from rural-based WWTPs to trimethoprim, cephalothin, penicillin, cotrimoxazole and streptomycin, respectively, positioning it as an emerging pathogen in the ECP of South Africa. Antimicrobial resistance has become a major medical and public health problem as it has direct links with disease management and containment. This is reflected by the increase in the fatality rate from 1 to 5.3 % after the emergence of drug resistance strains in Guinea-Bissau during the cholera epidemic of 1996–1997 (Dalsgaard et al. 2000).

Improved susceptibility to antibiotics was observed in *V. parahaemolyticus* which showed susceptibilities of ≥50 to 8 % of the test antibiotics and a resistance profile of ≥50 % to only 5 of the 13 test antibiotics (Fig. 5).

V. vulnificus showed a susceptibility profile of ≥50 to 7 % of the test antibiotics while showing a resistance profile of ≥50 to 6 % of the 13 antibiotics used (Fig. 6).

All three *Vibrio* species were susceptible to gentamycin, cefuroxime, meropenem and imipenem. Multiple antibiotic resistance patterns were also evident especially against such antibiotics as tetracycline, polymixin B, penicillin G, sulfamethoxazole and erythromycin against which all *Vibrio* species were resistant. Similar findings were reported by Keddy (2010a) during a cholera outbreak in South Africa in 2009, where the Enteric Diseases Research Unit (EDRU) at the National Institute for Communicable Diseases (NICD) processed 570 *V. cholerae* O1 isolates associated with the outbreak. Further laboratory characterisation of the isolates showed that they were 100 % resistant to co-trimoxazole, 48 % resistant to chloramphenicol, 100 % resistant to nalidixic acid, 3 % resistant to tetracycline and 39 % resistant to erythromycin.

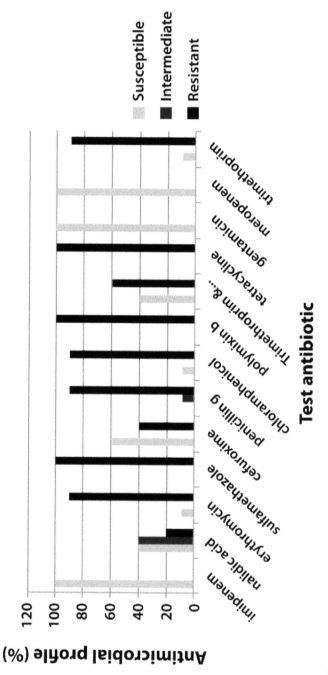

Figure 4. Antimicrobial profile for *V. fluvialis* (n=35).

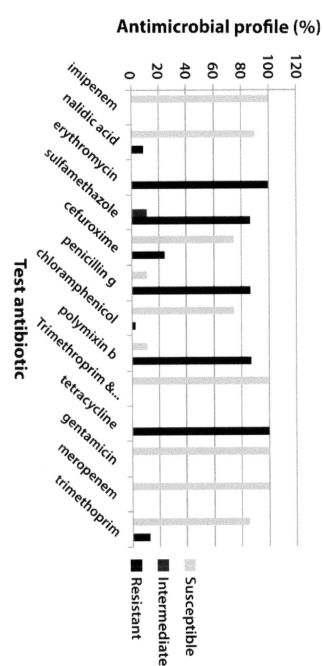

Figure 5. Antimicrobial profile of *V. parahaemolyticus* (n=86).

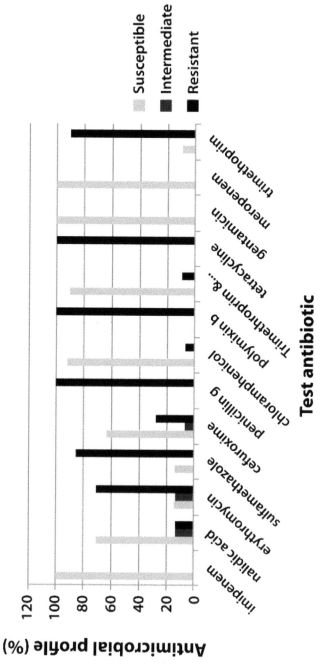

Figure 6. Antimicrobial profile of *V. vulnificus* (n=84).

In another outbreak in 2008, reported from Shebagold Mine in the Ehlanzeni district of Mpumalanga Province in South Africa, 31 isolates were submitted for analysis to the EDRU, revealing that all were biotype El Tor and displayed resistance to ampicillin, amoxycillin-clavulanate, sulfamethoxazole, trimethoprim, chloramphenicol, nalidixic acid, kanamycin, streptomycin and tetracycline, which was initially the antimicrobial agent of choice in the treatment of cholera in Africa, although they were susceptible to ciprofloxacin and imipenem (Keddy 2010b; Crowther-Gibson et al. 2011).

Our findings prove that wastewater effluents are an important source of antimicrobial resistant bacteria, as reported elsewhere (James et al. 2003; Byarugaba 2004). Atieno et al. (2013) also stated that the release of pathogenic enteric micro-organisms into aquatic environments can be a source of disease when water is used for drinking, recreational activities or irrigation. It has also been noted that the prevalence of pathogenic enteric bacteria in wastewater effluents (and hence in receiving water sources) increases public health risk if the bacteria are antibiotic-resistant because of the reduced efficacy of antibiotic treatment against human diseases caused by such bacteria (Tendencia and De la Pena 2002; Wenzel and Edmond 2009). Baine et al. (1977) reported that a large water-borne outbreak involving R+ bacteria (bacteria with R factors for antibiotic resistant gene transfer) led to a large number of deaths in Mexico, partly due to the failure of the patients to respond to antibiotics of choice. The *New York Times* (2013) quoted Centers for Disease Control (CDC) officials as having reported that at least 2 million Americans fall ill from antimicrobial-resistant bacteria every year and that at least 23,000 die from those infections. The paper reported that one particularly lethal type of drug-resistant bacteria, known as carbapenem-resistant Enterobacteriaceae (CRE), has become resistant to nearly all antimicrobials on the US market, further stating that though still relatively rare, CRE causes about 600 deaths a year in the US alone. Should the proliferation of antimicrobial resistant organisms be allowed to go unchecked, society will return to a time when people died from ordinary infections. This point is further buttressed by Torrice (Undated), in an article entitled "Multidrug Resistance Gene Released by Chinese WWTP", where he wrote:

In recent years, increasing numbers of patients worldwide have contracted severe bacterial infections that are untreatable by most available antibiotics. Some of the gravest of these infections are caused by bacteria carrying genes that confer resistance to a broad class of antibiotics called beta-lactams, many of which are treatments of last resort. Now a research team reports that some wastewater treatment plants in China discharge one of these potent resistance genes into the environment. Environmental and public health experts worry that this discharge could promote the spread of resistance.

There is also the possibility of antibiotic resistance genes being transmitted to autochthonous bacteria if such genes are carried by transferable and mobile genetic elements such as plasmids, thus contributing to the spread of antimicrobial resistance (Sayah et al. 2005). Development of drug resistance may be caused by the occurrence of antimicrobial agents at sub-optimal concentrations both in human bodies by continued usage and also in the wastewater matrix via leaching. The correlation between antimicrobial use and antibiotic resistance of commensal bacteria has been documented (Van den Bogaard and Stobberingh 2000). We infer, therefore, that the extent to which bacterial isolates are exposed to antibiotics before their release in the environment could be one of the reasons for the levels of antibiotic resistance shown by *Vibrio* isolates in this study. A lot has to be done, therefore, to prevent infections with multi-drug resistant organisms which find their way into the environment from WWTPs.

The public is endangered by exposure to wastewater, which happens directly or indirectly. Direct exposure routes include ingestion of contaminated water during recreational activities such as swimming, bathing and when undertaking religious ceremonies like baptisms. In the rural areas of most developing countries where the availability of piped water is limited and in most cases non-existent, the communities utilise stream/river water for drinking and other domestic uses. Indirect exposure routes include consumption of filter feeders such as molluscs which concentrate pathogenic microorganisms occurring in contaminated water (Tamburrini and Pozio 1999). In the face of climatic change and increasing water scarcity, wastewater is increasingly being considered as a new source of water for irrigation in regions where water is scarce (Blumenthal et al. 2000), exposing farmers to the risk of infection with *Vibrios* and other waterborne pathogens. Issues of public health, as related to both water quantity and quality, are, therefore, increasingly becoming of concern.

2.4 CONCLUSION

We conclude, therefore, that potentially virulent and multi-drug resistant *Vibrios* can be found in wastewater effluents of "healthy" communities. Even though *V. cholerae* was not assessed for in this work, we have reasonable suspicion, basing on the outcome of this work, that it is also prevalent in wastewater effluents owing to the existence of sub-clinical cholera cases in communities. The prevalence of *Vibrios* in the aquatic milieu constitutes a public health risk to people living in underdeveloped regions with no access to potable water. The use of surface water contaminated with wastewater effluents for either irrigation or recreational activities can likely pose risk of infection from waterborne pathogens. We recommend that future studies of this kind be directed at screening and characterising *V. cholerae* since it is the major pathogenic species under the genus *Vibrio* and more so since 31.8 % of the *Vibrio* isolates in this study fell outside the bracket of targeted *Vibrio* species.

REFERENCES

1. Abbott SL, Janda JM (1994) Severe gastroenteritis associated with Vibrio hollisae infection: report of 2 cases and review. Clin Infect Dis 18:310–312
2. Ahmadi A, Riahi H, Noori M (2005) Studies of the effects of environmental factors on the seasonal change of phytoplankton population in municipal waste water stabilization ponds. Toxicol Environ Chem 87:543–550
3. Arceivala SJ (1997) Sustainable wastewater treatment. AIC Watson Consultants Ltd., Mumbai
4. Asano T, Levine AD (1998) Wastewater reclamation, recycling, and reuse: an introduction. In: Asano T (ed) Wastewater reclamation and reuse. Technomic Publishing Company, Lancaster, pp 1–56

5. Atieno NR, Owuor OP, Omwoyo O (2013) Isolation of high antibiotic resistant fecal bacteria indicators, salmonella and Vibrio species from raw abattoirs sewage in peri-urban locations of Nairobi, Kenya. Greener J Biol Sci 3(5):172–178

6. Baine WB, Farmer JJ, Gangerosa EJ, Hermann GT, Thornsberry C, Rice PA (1977) Typhoid fever in the United States associated with the 1972–73 epidemic in Mexico. J Infect Dis 135:649–653

7. BLACKSASH (2010) Eastern Cape Provincial Health Consultative workshop: a community consultation workshop report 10–12 May 2010. Port Elizabeth, Eastern Cape Province. http://www.blacksash.org.za/files/ec_consultworkshp_10052010.pdf. Accessed on 26 June 2014

8. Blumenthal UJ, Mara DD, Peasey A, Ruiz-Palacios G, Stott R (2000) Guidelines for the microbiological quality of treated wastewater used in agriculture: recommendations for revising who guidelines. Bull World Health Organ 78(9):1104–1116

9. Bourne DE, Coetzee N (1996) An atlas of potentially water-related diseases in South Africa. WRC Report No 584/1/96, Pretoria, South Africa.

10. Byarugaba DK (2004) A view on antimicrobial resistance in developing countries and responsible risk factors. Int J Antimicrob Agents 24:105–110

11. Carnahan A, Harding MJ, Watsky D, Hansman S (1994) Identification of Vibrio hollisae associated with severe gastroenteritis after consumption of raw oysters. J Clin Microbiol 32:1805–1806

12. CDC (Centers for Disease Control and Prevention) (1999) Outbreak of Vibrio parahaemolyticus infection associated with eating raw oysters and clams harvested from Long Island Sound-Connecticut, New Jersey, and New York, 1998. Morb Mortal Wkly Rep 48:48–51

13. Centers for Disease Control and Prevention (2013) Cholera. Available at http://www.cdc.gov/cholera/index.html. Accessed 4 August, 2014.

14. Chakraborty R, Sinha S, Mukhopadhyay AK, Asakura M, Yamasaki S, Bhattacharya SK, Nair G, Ramamurthy T (2006) Species-specific identification of Vibrio fluvialis by PCR targeted to the conserved transcriptional activation and variable membrane tether regions of the toxR gene. J Med Microbiol 55:805–808

15. Chindah AC, Braide SA, Amakiri J, Izundu E (2007) Succession of phytoplankton in a municipal waste water treatment system under sunlight. Rev UDO Agríc 7:258–273

16. Crowther-Gibson P, Govender N, Lewis DA, Bamford C, Brink A, von Gottberg A, Klugman K, du Plessis M, Fali A, Harris B, Keddy KH, Botha M (2011) Part IV. GARP: human infections and antibiotic resistance. SAM J 101(8):567–578. doi:10.7196/samj.5102

17. Dalsgaard A, Forslund A, Serichantalergs O, Sandvang D (2000) Distribution and content of class 1 integrons in different Vibrio cholerae O-serotype strains isolated in Thailand. Antimicrob Agents Chemother 44:1315–1321

18. de Souza Costa Sobrinho P, Destro MT, Franco BDGM, Landgraf M (2010) Correlation between environmental factors and prevalence of Vibrio parahaemolyticus in oysters harvested in the Southern coastal area of Sao Paulo State, Brazil. Appl Environ Microb 76(4):1290–1293

19. DePola A, Nordstrom JL, Bowers JC, Wells JC, Cook DW (2003) Seasonal variation in the abundance of total and pathogenic Vibrio parahaemolyticus in Alabama oysters. Appl Environ Microbiol 69:1521–1526

20. Drury B, Rosi-Marshall E, Kelly JJ (2013) Wastewater treatment effluent reduces the abundance and diversity of benthic bacterial communities in urban and suburban rivers. Appl Environ Microbiol 79(6):1897–1905

21. ECSECC (2011) District demographic and socio-economic indicators: census 1996; 2001; 2011. http://www.ecsecc.org/files/library/documents/ECSECC_Census.pdf. Accessed 01 June 2014

22. Faruque SM, Islam MJ, Ahmad QS et al (2005) Self-limiting nature of seasonal cholera epidemics: role of host-mediated amplifi cation of phage. Proc Natl Acad Sci U S A 102:6119–24

23. Finkelstein R, Edelstein S, Mahamid G (2002) Fulminant wound infections due to Vibrio vulnificus. Isr Med Assoc J 4:654–655

24. Gerritsen J, Smidt H, Rijkers GT, de Vos WM (2011) Intestinal microbiota in human health and disease: the impact of probiotics. Genes Nutr 6:209–240. doi:10.1007/s12263-011-0229-7

25. Gilbert DN, Moellering RC, Sande MA (1999) Sanford guide to antimicrobial therapy, 29th edn. Antimicrobial Therapy, Inc., Hyde Park

26. Harris JB, LaRocque RC, Qadri F, Ryan ET, Calderwood SB (2012) Seminar: cholera. Lancet 379:2466–76

27. Heidelberg JF, Heidelberg KB, Colwell RR (2002) Bacteria of the gamma-subclass proteobacteria associated with zooplankton in chesapeake Bay. Appl Environ Microbiol 68:5498–5507

28. Heymann D (2008) Vibrio choleraeserogroups 01 and 0139. In: Control of communicable diseases manual. 19th ed. Washington, DC: Am Pub Health Ass 120–128

29. Farmer JJ (III), Hickman-Brenner FW (1992) The genera Vibrio and Photobacterium, p. 2952–3011. In Balows A, Trüper HG, Dworkin M, Harder W, Schleifer KH (ed.). The prokaryotes. A handbook on the biology of bacteria: ecophysiology, isolation, identification, and applications, 2nd ed. Springer-Verlag KG, Berlin, Germany

30. Hlady WG (1997) Vibrio infections associated with raw oyster consumption in Florida, 1981–1994. J Food Protect 60:353–357

31. Igbinosa EO, Okoh AI (2010) Vibrio fluvialis: an unusual enteric pathogen of increasing public health concern. Int J Environ Res Public Health 7:3628–3643

32. Igbinosa EO, Obi CL, Okoh AI (2009) Occurrence of potentially pathogenic Vibrios in the final effluents of a wastewater treatment facility in a rural community of the Eastern Cape Province of South Africa. Res Microbiol 160:531–537

33. Igbinosa EO, Obi CL, Okoh AI (2011) Seasonal abundance and distribution of Vibrio species in the treated effluents of wastewater treatment facilities in suburban and urban communities of Eastern Cape Province, South Africa. J Microbiol 49(2):224–232

34. Illinois Department of Natural Resources (2011) Illinois Coastal Management Program issue paper: Chicago River and North Shore Channel corridors. Illinois Department of Natural Resources. Springfield, IL

35. Iwamoto M, Ayers T, Mahon BE, Swerdlow DL (2010) Epidemiology of seafood-associated infections in the United States. Clin Microbiol Rev 23(2):399–411. doi:10.1128/CMR.00059-09

36. Jackson S. Beney C (2000) Detection of Vibrio cholerae in river water in the vicinity of an informal settlement. Presented at the WISA 2000 Biennial Conference, Sun City, South Africa, 28 May - 1 June 2000. http://www.ewisa.co.za/literature/files/237jackson.pdf. Accessed on 26 June 2014

37. James AE, Ian P, Helen S, Sojka RE (2003) Polyacrylamide+Al2(SO4)3 and polyacrylamide+CaO remove coliform bacteria and nutrients from swine wastewater. Environ Res 121:453–462

38. Janda JM, Powers C, Bryant RG, Abbott S (1988) Current perspectives on the epidemiology and pathogenesis of clinically significant Vibrio spp. Clin Microbiol Rev 1:245–267

39. Jiménez B, Barrios J, Mendez J, Diaz J (2004) Sustainable management of sludge in developing countries. Water Sci Technol 49(10):251–8

40. Keddy K (2010a) Cholera outbreak in South Africa: extended laboratory characterisation of isolates. In: National Health Laboratory Service—annual report 2009/2010. Sandringham, GA: National Health Laboratory Service.

41. Keddy K (2010b) Molecular characterisation of multidrug resistant cholera outbreak isolates. In: National Health Laboratory Service—annual report 2009/2010. Sandringham, GA: National Health Laboratory Service.

42. Kim YB, Okuda J, Matsumoto C, Takahashi N, Hashimoto S, Nishibichi M (1999) Identification of Vibrio parahaemolyticus strains at the species level by PCR targeted to the toxR gene. J Clin Microbiol 37:1173–1177

43. Kobayashi K, Ohnaka T (1989) Food poisoning due to newly recognized pathogens. Asian Med J 32:1–12

44. Kothary MH, Lowman H, Mccardell BA, Tall BD (2003) Purification and characterization of enterotoxigenic El Tor-like hemolysin produced by Vibrio fluvialis. Infect Immun 71:3213–20

45. Kwok AY, Wilson JT, Coulthart M, Ng LK, Mutharia L, Chow AW (2002) Phylogenetic study and identification of human pathogenic Vibrio species based on partial hsp60 gene sequences. Can J Micriobiol 48:903–910

46. Levine WC, Griffin PM (1993) Vibrio infections on the Gulf Coast: results of first year of regional surveillance Gulf Coast Vibrio Working Group. J Infect Dis 167:479–483

47. Mackintosh G, Colvin C (2003) Failure of rural schemes in South Africa to provide potable water. Environ Geol 44:101–105

48. Maugeri TL, Carbone M, Fera MT, Gugliandolo C (2006) Detection and differentiation of Vibrio vulnificus in seawater and plankton of coastal zone of the Mediterranean Sea. Res Microbiol 157:194–200

49. Mishra M, Mohammed F, Akulwar SL, Katkar VJ, Tankhiwale NS, Powar RM (2004) Re-emergence of El Tor Vibrio in outbreak of cholera in and around Nagpur. Indian J Med Res 120:478–480

50. Mohale NG (2003) Evaluation of the adequacy and efficiency of sewage treatment works in Eastern Cape. Rhodes University, South Africa, MSc. Thesis

51. Mukhopadhyay SK, Chattopadhyay B, Goswami AR, Chatterjee A (2007) Spatial variations in zooplankton diversity in waters contaminated with composite effluents. J Limnol 66:97–106

52. Nair L, Sudarsana J, Pushpa KK (2004) Epidemic of Salmonella enteritica serotype paratyphi a in Calicut-Kerala. Calicut Med J 2:e2

53. Nelson EJ, Chowdhury A, Flynn J et al (2008) Transmission of Vibrio cholerae is antagonized by lytic phage and entry into the aquatic environment. PLoS Pathog 4:e1000187

54. Nevondo TS, Cloete TE (2001) The global cholera pandemic. http://www.sciencein-africa.com/old/index.php?q=2001/september/cholera.htm. Accessed 04 June 2014

55. New York Times (2013) http://www.nytimes.com/2013/09/17/health/cdc-report-finds-23000-deaths-a-year-from-antibiotic-resistant-infections.html. Accessed 21 April 2014

56. Obi CL, Bessong PO, Momba MNB, Potegieter N, Samie A, Igumbor EO (2004) Profile of antibiotic susceptibilities of bacterial isolates and physicochemical quality of water supply in rural Venda communities of South Africa. Water SA 30:515–520

57. Okoh AI, Igbinosa EO (2010) Antibiotic susceptibility profiles of some Vibrio strains isolated from wastewater final effluents in a rural community of the Eastern Cape Province of South Africa. BMC Microbiol 10:143. doi:10.1186/1471-2180-10-143

58. Oliver JD, Kaper JB (2001) Vibrio species. In food microbiology 2nd edn (Doyle MP, Beuchat LR, Movtville TJ (Eds), pp 228–264 ASM Press, Washington. DC, USA

59. Osorio CR, Klose KE (2000) A region of the transmembrane regulatory protein ToxR that tethers the transcriptional activation domain to the cytoplasmic membrane displays wide divergence among Vibrio species. J Bacteriol 182:526–528

60. Pegram GC, Rollins N, Espey Q (1998) Estimating the cost of diarrhea and epidemic dysentery in KwaZulu-Natal and South Africa. Water SA 24(1):11–20

61. Sayah RS, Kaneene JB, Johnson Y, Miller RA (2005) Patterns of antimicrobial resistance observed in Escherichia coli isolates obtained from domestic- and wild-animal fecal samples, human septage, and surface water. Appl Environ Microbiol 71:1394–1404

62. Standard Methods (2005) Standard methods for the examination of water and waste-water. 20th Edn. American Public Health Association (APHA): Washington DC, USA.

63. Steinberg EB, Greene KD, Bopp CA, Cameron DN, Wells JG, Mintz ED (2001) Cholera in the United States, 1995–2000: trends at the end of the twentieth century. J Infect Dis 184(6):799–802

64. Tamburrini A, Pozio E (1999) Long-term survival of Cryptosporidium parvum oocysts in seawater and in experimentally infected mussels (Mytilus galloprovincialis). Int J Parasitol 29:711–715

65. Tarr CL, Patel JS, Puhr ND, Sowers EG, Bopp CA, Strockbine NA (2007) Identification of Vibrio isolates by a multiples PCR assay and rpoB sequence determination. J Clin Microbiol 45(1):134–140

66. Tendencia EA, De la Pena LD (2002) Level and percentage recovery of resistance to oxytetracycline and oxolinic acid of bacteria from shrimp ponds. Aquaculture 213:1–13

67. Todar K (2005) http://textbookofbacteriology.net/cholera.html Accessed 06 June 2014.
68. Torrice M (Undated) Multidrug resistance gene released by Chinese wastewater treatment plants. http://science-beta.slashdot.org/story/13/12/18/0013200/multi-drug-resistance-gene-released-by-chinese-wastewater-treatment-plants. Accessed on 08 April 2014
69. Van den Bogaard AE, Stobberingh EE (2000) Epidemiology of resistance to anti-biotics—links between animals and humans. Int J Antimicrob Agents 14:327–335
70. WASH (2010) News about water, sanitation and hygiene (WASH) in Africa. URL: http://washafrica.wordpress.com/2010/07/16/south-africa-shortage-of-fresh-water-supplies-looming/. Accessed on 30 April 2014
71. Wen Q, Tutuka C, Keegan A, Jin B (2009) Fate of pathogenic microorganisms and indicators in secondary activated sludge wastewater treatment plants. J Environ Manag 90:1442–1447
72. Wenzel RP, Edmond MB (2009) Managing antibiotic resistance. J Med 343:1961–1963
73. Wong RS, Chow AW (2002) Identification of enteric pathogens by heat shock protein 60 kDa (HSP60) gene sequesnces. FEMS Microbiol Lett 206:107–113

CHAPTER 3

Performance of the Municipal Wastewater Treatment Plant for Removal of *Listeria monocytogenes*

NAHID NAVIDJOUY, MOHAMMAD JALALI, HOSSEIN MOVAHEDIAN ATTAR, AND HAJAR AGHILI

3.1 BACKGROUND

Listeriosis is essentially a foodborne disease caused by *L. monocytogenes* and to some extent *L. ivanovii*. The disease condition vary from that affect immunocompromised patients to febrile gastroenteritis and prenatal infections associated with fetal loss or abortion in humans and animals. [1] Although rare, the disease reported to have very high mortality rare (20-50%), thus making it of serious public health concern.[2]

L. monocytogenes commonly is a saprophytic organism living naturally in the plant-soil environment. The ubiquitous nature of *L. monocytogenes* results in contamination of numerous food products including meat, milk and dairy products, sea foods and vegetables.[3] A possible agricultural route of human exposure to *Listeria* is through the ingestion of uncooked food crops grown in the lands irrigated with contaminated water and/or fertilized with *Listeria*-contaminated biosolids.[4]

L. monocytogenes has been detected in wastewater,[5-12] sewage sludge[5,7,13-15] and *L. innocua* has been detected in compost and irrigation water.[16] Sewage sludge can be regarded as a reservoir of *L. monocytogenes* and the presence of this pathogenic bacterium in such fertilizer would increase the risk of crops contamination. Therefore, biological monitoring of the presence of pathogenic bacteria in the effluent and sludge of sewage treatment plants is of special importance.[14] In addition, numerous rules and regulations ratified for the existence of certain microorganism in effluent and sludge. For instance, US Environmental Protection Agency (EPA) has been setup an extensive regulation for reuse of effluent and sludge in agriculture lands.[17] However, recent studies have been indicated that the traditional standards are not effective on environmental dissemination of new emerging microorganisms.[18] In the other hand with reports of inadequate removal of *Listeria* from wastewater in developed world,[19,20] one can safely presume that wastewater treatment plants in Iran are insufficient at removing these pathogens from wastewater influents prior to reused for irrigating the farmlands.

The occurrences of *L. monocytogenes* in variety of foods including vegetables have been reported in Iran.[21,22] However, to our knowledge there are no study on the prevalence of *L. monocytogenes* and its removal by sewage treatment plants in Iran. Therefore, the present study examined the prevalence of *Listeria* spp. in various wastewater and sludge samples in Northern municipal wastewater treatment plant in Isfahan, Iran. We also evaluated the performance of this treatment plant on removal *Listeria* spp.

3.2 MATERIALS AND METHODS

3.2.1 PLANT DESCRIPTION

The current investigation was conducted at the Northern municipal wastewater treatment plant in Isfahan, Iran. The Isfahan city has three large wastewater treatment plants in North, South and East with 1,200,000,

800,000 and 250,000 people of nominal capacity, respectively. The Northern treatment plant is the largest one in which treated a part of sewage of the city with the inlet flow rate of 1.6 m³/sec currently. The flow diagram of the plant is shown in Figure 1.

3.2.2 SAMPLE COLLECTION

Samples were collected on a monthly basis from the influent, effluent, raw sludge, stabilized sludge and dried sludge between December 2010 and January 2011. As shown in Figure 1, various samples were collected from five spots of this sewage treatment plant based on the standard method. [23] Each site was sampled 13 times and in total 65 samples obtained. Sewage and sludge samples collected in the plastic containers (500 ml) and plastic bags, respectively and transported in cooler boxes containing ice packs to the School of Health Laboratory, Isfahan University of Medical Sciences for analysis.

3.2.3 SAMPLE ANALYSIS

Samples were analyzed for the presence of *Listeria* spp. using cultivation methodology recommended by United States Department of Agriculture (USDA).[23] Briefly, a 10g of sample was transferred to 90 ml of first enrichment medium UVM I (University of Vermont Media formula) (Merck, Germany) incubated in 30°C for 24 h. 0.1 ml of primary enrichments were transferred to 10 ml of UVM II (Fraser broth) (Merck, Germany) and incubated at 35-37°C for 48 h. Secondary enrichments were streaked on PAL-CAM agar supplemented with selective supplement HC784958 (Merck, Germany). The plates were examined for typical *Listeria* colonies. At least 3 to 5 typical colonies were selected and purified on the TSAYE culture medium. All isolates were subjected to standard biochemical tests such as Gram staining, carbohydrate fermentation (L-rhamnose, D-xylose, glucose, D-manitol), catalase test, MR/VP test, motility test at 25°C and CAMP test.[24,25]

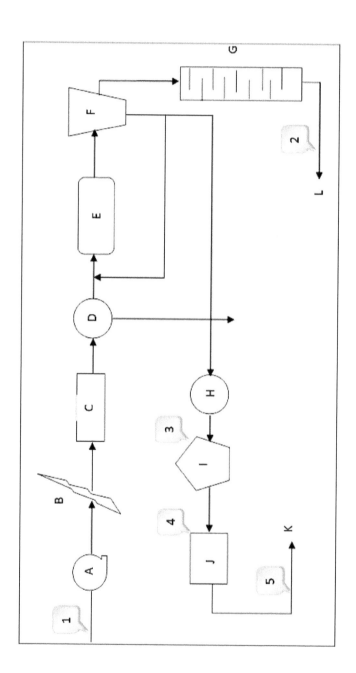

Figure 1. Flow diagram of Northern Isfahan, Iran sewage treatment plant, A: Pumping station, B: Screening, C: grit removal, D: Primary sedimentation tank, E: Aeration tank, F: secondary sedimentation tank, G: Chlorination basin, H: Thickening sludge, I: Anaerobic digester, J: Sludge drying bed, K: Amendment soil, L: Effluent. The numbers indicate the sampling spots. 1) Influent, 2) Effluent, 3) raw sludge (primary mixed sludge + secondary sludge), 4) stabilized sludge (raw sludge processed by the digester), 5) dried (final) sludge.

In order to quantified *L. monocytogenes*, 3-tube most probable number (MPN) method was used in supplemented Fraser broth culture medium. The wastewater and sludge samples were cultivated by the amounts of 10 ml, 1 ml and 0.1 ml in each tube.[14,20] The samples of raw sludge, stabilized sludge and the dried sludge were diluted up to 10^{-1} and 10^{-2} respectively and MPN of samples was determined and the dilution coefficients were taken into consideration.

The positive tubes (blackened) were confirmed using biochemical and PCR methods.[25] The pair of primers 234 (5'-CATCGACGGCAACCTC-GGAGA-3') and primer 319 (5'-ATCAATTACCGTTCTCCACCATT-3') were used to amplify a 417 bp internal fragment of the blyA gene.[26] PCR was achieved as described by Fitter et al., and 16 µl of the PCR amplified reaction mixtures were subjected to horizontal gel electrophoresis in 1.8% agarose gels run in 1× Tris-borate (TBE). PCR products were visualized using ethidium bromide staining (1 µg/ml) and photographed under UV light. *L. monocytogenes* serotype 4a (IRTCC 1293) provided from Razi Institute in Iran used as a positive control.

3.2.4 STATISTICAL ANALYSIS

Calculation of mean and standard deviations were performed using Microsoft Excel office 2007 version. Test of significance (independent T-test) were performed using SPSS 10 version. Independent t-test was used to compare differences in the *L. monocytogenes* MPN mean between influent and effluent, raw sludge and stabilized sludge, and stabilized sludge and dried sludge. All tests of significance were considered statistically significant at P value ≤ 0.05.

3.3 RESULTS

The prevalence of *Listeria* spp. in different samples collected in various stages of Northern Isfahan's sewage treatment plant is shown in Figure 2. In the current study 52/65 (80%) of all samples were positive for *Listeria* spp. and, *L. monocytogenes*, *L. innocua*, *L. seeligeri* were isolated from

41/65 (63.1%), 30/65 (44.6%), 10/65 (15.4%) of the tested samples, respectively. The mean MPN of *Listeria* spp. in different sewage samples are shown in and Table 1. The independent t-test showed a significant difference in mean of MPN of *L. monocytogenes* between the influent and effluent samples (P value = 0.036), the raw sludge and stabilized sludge (P value = 0.025), and also the stabilized sludge and the dried sludge (P value = 0.027). The removal percentage of *L. monocytogenes* in sewage treatment process, the digester tank, and sludge drying media was determined as 69.6%, 64.7%, and 73.4%, respectively. All *L. monocytogenes* isolates also confirmed by PCR method and amplified a 417 bp internal fragment of the blyA gene [Figure 2].

3.4 DISCUSSIONS

Listeria monocytogenes is considered a ubiquitous foodborne pathogen which can lead serious infections in human. One of the potential sources of this bacterium is municipal wastewater treatment plants. Sludge and effluent of sewage treatment plants are using in agricultural farms as a fertilizer in Iran. Therefore, the bacteria may spread on agriculture land and other environment. This may in particular, contaminate the foods of plant origin and cause a risk of spreading listeriosis to human and animals. Despite the isolation of *Listeria* spp. from various foods and vegetables in Iran,[21] there are no data on occurrence of bacterium in municipal sewage so far. Therefore, this study was conducted to isolate and enumerate *L. monocytogenes* in the sewage treatment plant in Northern Isfahan, Iran. The provided data can be used to develop appropriate control strategies of disease in human and animal.

In the current study 80% of all samples were positive for *Listeria* spp. and, *L. monocytogenes*, *L. innocua*, *L. seeligeri* were isolated from 63.1%, 44.6% and 15.4% of the tested samples, respectively. *L. monocytogenes*, *L. innocua* and L. seeligeri were isolated from 76.9%, 23.1% and 23.1% of influent, 38.5%, 46.2% and 7.7% of effluent, 84.6%, 69.2% and 46.2% of raw sludge, 69.2%, 76.9% and 0% of stabilized sludge and 46.2%, 7.7% and 0% of dried sludge samples, respectively. Considering a long detention time (20 days for anaerobic digesters and 2-3 month for sludge drying

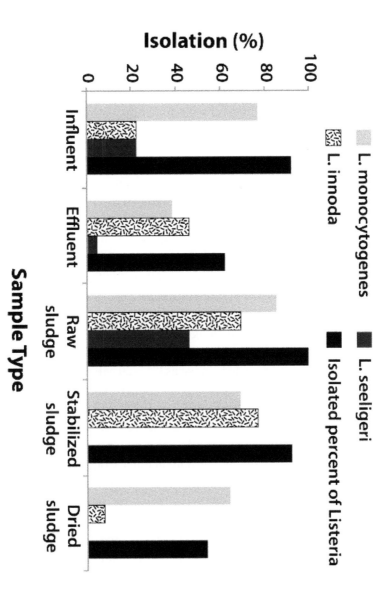

Figure 2. Percentage of *Listerias* spp. isolation in various samples collected from different spots of Northern Isfahan's, Iran Sewage Treatment.

beds) and suitable temperature for the growth of mesophilic bacteria in sludge digesters (In order to stabilize the organic material, reduce the volatile organics and pathogenic bacteria population and improvement of the sludge dewatering characteristics) the above results show that *L. monocytogenes* was dominant species in all samples whereas, *L. innocua* was more common in effluent samples. Also, *L. seeligeri* was not found in the raw and stabilized sludge and all isolated species from dried sludge were (except one for *L. innocua*) *L. monocytogenes*. These findings indicate that other species are more sensitive in compared to *L. monocytogenes* to the environmental factors during treating process. This is in agreement with the report of Spanish study[15] that indicated *L. seeligeri* and *L. innocua* were more destructed during sewage treatment. It seems that the pathogenic species of *L. monocytogenes* was able to survive even after digestion and sludge drying beds. That may indicate the importance of monitoring of this species in the sludge and sewage treatment plants.

Similar to our finding, most of the environmental study reported high rate of contamination. For example, a study in 2005 in France,[20] isolated *Listeria* spp. from 84.4% of effluent and 89.2% raw sludge. In other study, Garrec et al., (2003) also isolated *Listeria* spp. and *L. monocytogenes* in 87% and 73% of dewatered sludge and in 96% and 80% of sludge stored in tanks respectively. They found the number of *L. monocytogenes* in dewatering sludge and digested sludge as 0.15-20 MPN/g and 1-240 MPN/g, respectively.[14] This result is slightly higher than what has been found in current study. Also Odjajar et al., in South Africa reported the detection of free-living *Listeria* species and associated-planktons in 96% and 58-67% of final effluents of a rural wastewater treatment facility samples, respectively.[12] This research group, in other work reported that the *Listeria* population of 2.9×10^0 to 1.2×10^5 cfu/ml in municipal wastewater effluents.[27]

A study in Poland, reported the isolation rate of *Listeria* in mechanical and biological treatment plant samples as 90% *L. monocytogenes*, 5% L. seeligeri and 5% L. grayi from the point of municipal sewage delivery to the treatment plant.[19] Also a study in Sweden reported significantly lower rate of isolation of *L. monocytogenes* from 8/64 (12%) of collected raw sludge samples and 1/69 (2%) from treated sludge samples.[28]

TABLE 1. Mean MPN of *Listeria* spp. in the sewage and sludge of northern isfahan's sewage treatment plant.

Listeria spp.	Sample type	Unit	Mean MPN	Maximum	Standard deviation
L. monocyto-genes	Influent	MPN/100mL	6.08	28	7.5
	Effluent	MPN/100mL	1.85	11	3.16
	Raw sludge	MPN/g dry matter	57	233	60
	stabilized sludge	MPN/g dry matter	20.12	88.46	24.2
	Dried sludge (Final)	MPN/g dry matter	5.53	23.9	7.3
L. innocua	Influent	MPN/100mL	1	6	2
	Effluent	MPN/100mL	3.92	15	6.6
	Raw sludge	MPN/g dry matter	25	58.3	21.16
	stabilized sludge	MPN/g dry matter	15.7	42.3	13.55
	Dried sludge (Final)	MPN/g dry matter	0.5	6.52	1.8
L. seeligeri	Influent	MPN/100mL	0.69	3	1.32
	Effluent	MPN/100mL	0.46	6	1.66
	Raw sludge	MPN/g dry matter	14.75	58.33	18.68
	stabilized sludge	MPN/g dry matter	0	0	0
	Dried sludge (Final)	MPN/g dry matter	0	0	0

In the present work, the efficiency of wastewater treatment processes, digester tank and drying bed in removal of *L. monocytogenes* were 69.6%, 64.7% and 73.4%, respectively. In contrast, a study in Spain, reported the higher average efficiency of 92% removal of *Listeria* species in the sewage treatment plant.[15]

Alghazali et al., in Iraq, reported higher prevalence of *Listeria* spp. of 100% in treated wastewater effluent but lower densities of 3-28 MPN/ml compare to this study. They have also reported higher efficiency rate of a municipal sewage treatment plant of removal of 85-99.7% for *Listeria* species.[7] Similar to the current study, Paillard et al., reported 84.4% prevalence of *Listeria* spp. in treated wastewater in France with population ranging from 0.3 to 21 MPN/ml.[20] Also, in Holland *L. monocytogenes* isolated from 90% and *L. seeligeri* and *L. graee* both from 5% of income untreated sewage.[19]

The number of biochemical test which has been used to differentiate species within the genus *Listeria*, is limited.[29,30] The main distinctive criteria between *L. monocytogenes* and *L. innocua* based on the haemolytic activity of the former is CAMP test.[31] However, this test is not reliable method for characterization *Listeria* species and distinction between *L. monocytogenes* and *L. innocua*.[32] Therefore, in this study, all of *L. monocytogenes* isolates also confirmed by PCR method [Figure 3].

In the current study, the *L. monocytogenes* species was detected as a dominant species in all samples of sewage and sludges except in the final effluent and stabilized (that *L. innacua* was dominant species). This finding is in agreement with the most of the previous studies. The raw sludge had the highest MPN of *L. monocytogenes*. This result indicated that *L.*

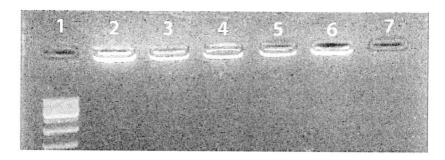

Figure 3. Agarose gel showing the amplified PCR product of *L. monocytogenes* isolated from sewage and sludge samples; Lane 1:100bp DNA ladder; Lane 2: positive control, *L. monocytogenes* serotype 4a (IRTCC 1293) provided from Razi Institute in Iran; Lane 3: *L. monocytogenes* isolated from the effluent; Lane 4: *L. monocytogenes* isolated from raw sludge; Lane 5: *L. monocytogenes* isolated from the stabilized sludge; Lane 6: negative control.

monocytogenes is more resistant compare to other *Listeria* species. As it was survived after the digestion process and sludge drying bed. The efficiency of the Northern Isfahan's Sewage Treatment Plant to remove the bacterium in the effluent and dried sludge was 69.6% and 90.2%, respectively. As shown in Figure 2, *L. monocytogenes* isolated from 38.5% and 46.2% of the final effluent and sludge samples, respectively. These results have shown that *L. monocytogenes* were present in wastewater treatment plant effluents and sludge at high level. There are no regulation on reuse of effluent for irrigation of farmlands and the dried sludge as a fertilizer in Iran concerning *L. monocytogenes*, so the *L. monocytogenes* population may increase in soil, surface waters, and the nature. In this respect, the reuse of effluent and sludge may spread the bacterium in the environment and infect humans and animals. Therefore, it is recommended to monitor regularly *L. monocytogenes* contamination in soil and the agricultural products of the lands irrigated with effluent of the treatment plant. In addition, the use of sewage sludge as a fertilizer need to be precisely reconsiders by regulatory authorities.

REFERENCES

1. Siegman-Igra Y, Levin R, Weinberger M, Golan Y, Schwartz D, Samra Z, et al. Listeria monocytogenes infection in Israel and review of cases worldwide. Emerg Infect Dis 2002;8:305-310.
2. Lyautey E, Lapen DR, Wilkes G, McCleary K, Pagotto F, Tyler K, et al. Distribution and characteristics of Listeria monocytogenes isolates from surface waters of the South Nation River watershed, Ontario, Canada. Appl Environ Microbiol 2007;73:5401-5410.
3. Sant'Ana AS, Igarashi MC, Landgraf M, Destro MT, Franco BD. Prevalence, populations and pheno- and genotypic characteristics of Listeria monocytogenes isolated from ready-to-eat vegetables marketed in São Paulo, Brazil. Int J Food Microbiol 2012;155:1-9.
4. Pachepsky Y, Shelton DR, McLain JE, Patel J, Mandrell RE. Chapter Two-Irrigation Waters as a Source of Pathogenic Microorganisms in Produce: A Review. In: Donald LS, editor. Advances in Agronomy. United States: Academic Press; 2011. p. 75-141.
5. Watkins J, Sleath KP. Isolation and enumeration of Listeria monocytogenes from Sewage, Sewage Sludge and River Water. J Appl Bacteriol 1981;50:1-9.
6. Geuenich HH, Muller HE, Schretten-Brunner A, Seeliger HP. The occurrence of different Listeria species in municipal waste water. Zentralbl Bakteriol Mikrobiol Hyg B 1985;181:563-565.

62   Wastewater and Public Health: Bacterial and Pharmaceutical Exposures

7. Al-ghazali M, Al-azawi SK. Detection and enumeration of Listeria monocytogenes in a sewage treatment plant in Iraq. J Appl Microbiol 1986;60:251-254.

8. Jepsen SE, Krause M, Grüttner H. Reduction of fecal Streptococcus and Salmonella by selected treatment methods for sludge and organic waste. Water Sci Technol 1997;36:203-310.

9. Dumontet S, Dinel A, Baloda S, editors. Pathogen reduction in biosolids by composting and other biological treatments: A literature review. International Congress, Maratea; 1997.

10. Stampi S, De Luca G, Varoli O, Zanetti F. Occurrence, removal and seasonal variation of thermophilic campylobacters and Arcobacter in sewage sludge. Zentralbl Hyg Umweltmed 1999;202-219.

11. Welshimer HJ, Donker-Voet J. Listeria monocytogenes in nature. Appl Microbiol 1971;21:516-516.

12. Odjadjare EE, Okoh AI. Prevalence and distribution of Listeria pathogens in the final effluents of a rural wastewater treatment facility in the Eastern Cape Province of South Africa. World Microbiol Biotechnol 2010;26:297-307.

13. De Luca G, Zanetti F, Fateh-Moghadm P, Stampi S. Occurrence of Listeria monocytogenes in sewage sludge. Zentralbl Hyg Umweltmed 1998;201:269.

14. Garrec N, Picard-Bonnaud F, Pourcher AM. Occurrence of Listeria spp and L. monocytogenes in sewage sludge used for land application: Effect of dewatering, liming and storage tank on survival of Listeria species. Fems Immunol Med Microbiol 2003;35:275-283.

15. Combarro M, Gonzalez M, Araujo M, Amezaga A, Sueiro R, Garrido M. Listeria species incidence and characterization in a river reciving town sewage from a sewage treatment plant. J Water 1997;35:201-204.

16. Moreno Y, Ballesteros L, Garcia-Hernandez J, Santiago P, Gonzalez A, Antonia Ferrus M. Specific detection of viable Listeria monocytogenes in Spanish wastewater treatment plants by Fluorescent In Situ Hybridization and PCR. Water Res 2011;45:4634-4640.

17. EPA. Environment regulations Technology, control of pathogens and vector attraction in sewage sludge. Avialable from: http://water.epa.gov/ scitech/wastetech/biosolids/index.cfm.

18. Toxicants NRCCo, Land PiBAt. Biosolids applied to land: Advancing standards and practices: Natl Academy Pr; 2002.

19. Czeszejko K, Bogusławska-Was E, Dabrowski W, Kaban S, Umanski R. Prevalence of Listeria monocytogenes in municipal and industrial sewage. Electron J Pol Agric Univ Environ Dev 2003;6:1-8.

20. Paillard D, Dubois V, Thiebaut R, Nathier F, Hoogland E, Caumette P, et al. Occurrence of Listeria spp. in effluents of French urban wastewater treatment plants. Appl Environ Microbiol 2005;71:7562-7566.

21. Jalali M, Abedi D. Prevalence of Listeria species in food products in Isfahan, Iran. Int J Food Microbiol 2008;122:336-340.

22. Rahimi E, Ameri M, Momtaz H. Prevalence and antimicrobial resistance of Listeria species isolated from milk and dairy products in Iran. Food Control 2010;21:1448-1452.

23. A.P.H.A. standard method for examination of water and wastewater. Washington DC: American Public Health Association; 2003.
24. Harrigan W. Labratory Methods in food micribiology 3rd ed. United States: Academic Press 1998.
25. Bridson E. The oxoid manual unipath ltd. 9th ed. 2006. Avilable from: http://ebook-browse.com/oxoid-manual-9th-edition-pdf-d65014325: [Last accessed on 2006].
26. Fitter S, Heuzenroeder M, Thomas C. A combined PCR and selective enrichment method for rapid detection of Listeria monocytogenes. J Appl Microbiol 1992;73:53-59.
27. Odjadjare EE, Obi LC, Okoh AI. Municipal wastewater effluents as a source of Listerial pathogens in the aquatic milieu of the Eastern Cape Province of South Africa: A concern of public health importance. Int J Environ Res Public Health 2010;7:2376-2394.
28. Sahlström L, Aspan A, Bagge E, Tham MLD, Albihn A. Bacterial pathogen incidences in sludge from Swedish sewage treatment plants. Water Res 2004;38:1989-1994.
29. Garrec N, Sutra L, Picard F, Pourcher AM. Development of a protocol for the isolation of Listeria monocytogenes from sludge. Water Res 2003;37:4810-4810.
30. Rocourt J, Seeliger HP. Distribution of species of the genus Listeria]. Zentralbl Bakteriol Mikrobiol Hyg A 1985;259:317-330.
31. McKellar R. Use of the CAMP test for identification of Listeria monocytogenes. Appl Environ Microbiol 1994;60:4219.
32. Bille J, Catimel B, Bannerman E, Jacquet C, Yersin M, Caniaux I, et al. API Listeria, a new and promising one-day system to identify Listeria isolates. Appl Environ Microbiol 1992;58:1857-1860.

PART II

ANTIBIOTIC RESISTANCE

Antibiotic Resistance and Prevalence of Class 1 and 2 Integrons in *Escherichia coli* Isolated from Two Wastewater Treatment Plants, and Their Receiving Waters (Gulf of Gdansk, Baltic Sea, Poland)

EWA KOTLARSKA, ANETA ŁUCZKIEWICZ, MARTA PISOWACKA, AND ARTUR BURZYŃSKI

4.1 INTRODUCTION

Safe and economical way of wastewater disposal is an important problem requiring proper receiver-oriented management. Nowadays, wastewater treatment focuses mainly on parameters that may cause oxygen depletion and eutrophication of the receiving waters: organic matter, nitrogen, and phosphorus. Health aspects are considered only in terms of fecal contamination and evaluated only in bathing areas by monitoring fecal indicators (*Escherichia coli* and *Enterococcus* species). However, other important

© *The Author(s) 2014.* Environmental Science and Pollution Research, *August 2014; 10.1007/s11356-014-3474-7. This is an open access article distributed under the terms of the Creative Commons Attribution License.*

aspects of wastewater discharge are currently under debate. It is suspected that clinically relevant bacteria and mobile genetic elements can survive the wastewater treatment plant processes (Reinthaler et al. 2003; D'Costa et al. 2006; Łuczkiewicz et al. 2010) and be disseminated in the receiving waters (Iwane et al. 2001; Li et al. 2009; Czekalski et al. 2012). Additionally, human-associated bacteria are regarded as important vectors of gene transmission (D'Costa et al. 2006). Thus, domestic and municipal wastewater should be considered in global antibiotic resistance gene dissemination.

Mobile genetic elements play crucial role in spreading antimicrobial-resistance genes. Among them, integrons are suspected to be the most important (Stalder et al. 2012), since they are often associated with other mobile genetic elements, such as plasmids or transposons, and are detected in various environments and matrices (Rosser and Young 1999; Goldstein et al. 2001; Elsaied et al. 2007, 2011; Gillings et al. 2009; Xia et al. 2013). Integrons are widely distributed in gram-negative bacteria; however, recent studies indicate their presence also in gram-positive species (Xu et al. 2001, 2010). About 10 % of bacterial genomes that have been partially or completely sequenced harbor this genetic element (Boucher et al. 2007). The spread of antimicrobial-resistance genes by integrons was extensively investigated primarily in clinical settings (Grape et al. 2003; Pan et al. 2006; Dubois et al. 2007). Nowadays, increased interest in their environmental role is observed (Elsaied et al. 2007; Gillings et al. 2009; Elsaied et al. 2011) and also reviewed by Stalder et al. (2012). Integrons consist of integrase (intI) gene, a recombination site (attI) and one or two promoters (Recchia and Hall 1995). Based on the amino acid sequence of the IntI protein, five classes of integrons have been described (Cambray et al. 2010). Classes 1, 2, and 3 are the most common. Gene cassettes are captured through recombination between attI site of the integron and a 59-bp element (attC site) from the cassette. Gene cassettes associated to class 1–3 integrons often confer resistance to aminoglycoside and β-lactam antibiotics, chloramphenicol, trimetophrim, sulfonamides, spectinomycin and others (Partridge et al. 2009; Moura et al. 2009). They can also carry other adaptive genes associated with environmental stresses or open reading frames, coding hypothetical proteins of unknown function (Elsaied et al. 2007, 2011).

The objective of this study was to investigate antibiotic-resistance pro-files in *E. coli* isolated from two local wastewater treatment plants (raw and treated wastewater samples), their marine outfalls located in the Gulf of Gdansk, the Baltic Sea (Poland), and from major tributary of the Baltic Sea—the Vistula River. In order to evaluate the role of the studied waste-water effluents and tributaries in dissemination of integrons and antibiotic resistance genes in anthropogenically impacted part of the Gulf of Gdansk, prevalence of class 1 and 2 integrons among *E. coli* isolates resistant to at least one antimicrobial agent was analyzed. The association between resis-tance or multiresistance to tested antimicrobials and presence of integrons in *E. coli* isolates was also studied. To assess the diversity of gene cassettes in integron-positive isolates, 35 selected amplicons representing variable region of integrons were sequenced and annotated.

4.2 MATERIALS AND METHODS

4.2.1 SAMPLING SITES AND SAMPLES COLLECTION

Altogether, 36 samples were analyzed. Samples of raw and treated waste-water (RW and TW, respectively) were taken from two local wastewater treatment plants (WWTPs) and from their marine outfalls (MOut) (Fig. 1). In both cases, treated wastewater is discharged by submarine collec-tors, about 2.5 km long and equipped with diffuser systems. The plants mainly treat municipal wastewater. Industrial wastewater and non-disin-fected hospital wastewater consist about 10 and 0.2 % of their daily in-flow, respectively.

4.2.2 WWTP "GDANSK–WSCHOD"

WWTP Gdansk–Wschod works in a modified University of Cape Town (UCT) type system, with integrated effective removal of nitrogen, phos-phorus, and carbon in anaerobic/anoxic/oxic zones fed with internal re-cycles (Tchobanoglous et al. 1991). The plant serves the population of about 570,000 people (population equivalent—700,000; the average daily

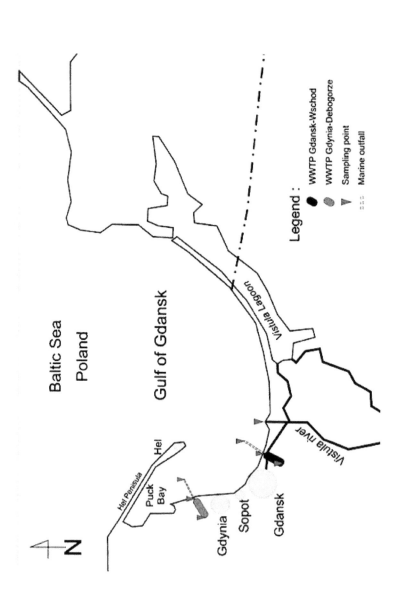

Figure 1. Sampling sites: raw (WRW) and treated (WTW) wastewater, as well as marine outfall (WMOut) of WWTP Gdansk–Wschod; raw (DRW) and treated (DTW) wastewater, as well as marine outfall (DMOut) of WWTP Gdynia–Debogorze. VR mouth of the Vistula River

flow—96,000 m³/day). Marine outfall has been operated since 2001. The flow-proportioned composite samples of raw (WRW) and treated wastewater (WTW) were taken from February to December 2011 (six samples of each type of wastewater were collected). Marine outfall (WMOut) sampling point was located in the Gulf of Gdańsk (54° 22' 44.4" N; 18° 52' 40.8" E), and marine water was taken in April, June, August, and October 2011 (four samples in total were collected).

4.2.3 WWTP "GDYNIA–DEBOGORZE"

WWTP Gdynia–Debogorze treats wastewater in four-stage Bardenpho system at the biological stage of the plant, including primary and secondary anoxic reactors (Tchobanoglous et al. 1991). The average daily flow was 55,000 m3/day, population equivalent—440,000. Treated wastewater has been discharged there by marine outfall since 2010. Raw (DRW) and treated wastewater (DTW) were taken as flow proportioned samples every 2 months from February to December 2011 (six samples of each type of wastewater were collected). Marine water samples (DMOut) were collected in the Puck Bay (54° 37' 08.4" N; 18° 33' 28.8" E), in April, June, August, and October 2011 (four samples in total were collected).

4.2.4 MOUTH OF THE VISTULA RIVER

Mouth of the Vistula River (VR) was sampled at the point located at 54° 22' 58.8" N; 18° 58' 04.8" E. Samples of surface water were taken once a month in April, June, August, and October 2011 (four samples in total were taken).

4.2.5 ENUMERATION AND ISOLATION OF E. COLI

E. coli detection in wastewater and marine water samples was carried out according to APHA (1998) by means of membrane filtration, using membrane fecal coliform (mFC) agar (Merck) and an appropriate dilution (treated wastewater) or concentration (marine and riverine water) of collected samples. The enumeration was conducted for wastewater samples in triplicate (according to recommendations of APHA (1998) for water and wastewater samples), and for other samples, due to suspected low number of *E. coli*, in sixfold repetition. For quality control, *E. coli* ATCC 25922 strain was used. This strain forms blue and dark blue colonies on mFC agar while other organisms form gray to cream colonies. For identification and antimicrobial susceptibility tests, 5–15 blue bacterial colonies (presumptive *E. coli*) from each of the membrane filters were selected and preserved at minus 80 °C in nutrient broth (beef extract, 3 g/L, peptone 5 g/L; Becton, Dickinson and Company), supplemented with 15 % glycerol.

4.2.6 IDENTIFICATION AND ANTIMICROBIAL SUSCEPTIBILITY TESTS

The identification and drug susceptibility of presumptive *E. coli* isolates were determined by the Phoenix Automated Microbiology System (Phoenix AMS, BD) according to the manufacturer's instruction. The susceptibility tests, based on microdilution, were carried out against 17 antimicrobial agents: amikacin (AN), gentamicin (GM), tobramycin (NN), imipenem (IPM), meropenem (MEM), cefazolin (CZ), cefuroxime (CXM), ceftazidime (CAZ), cefotaxime (CTX), cefepime (FEP), aztreonam (ATM), ampicillin (AM), amoxicillin/clavulanate (AMC), piperacillin/tazobactam (TZP), trimethoprim/sulfamethoxazole (STX), ciprofloxacin (CIP), and levofloxacin (LVX) together with screening for extended-spectrum β-lactamases (ESBL) production. Antimicrobial susceptibility was categorized according to EUCAST (2011). All ESBL-producing isolates were confirmed using the double-disk synergy test according to EUCAST (2011). The multidrug-resistance (MDR) phenotype was defined as simultaneous resistance to antimicrobial agents representing 3 or more categories, according to Magiorakos et al. (2012). *E. coli* ATCC 25922 and Pseudomonas aeruginosa ATCC 27853 were used as reference strains. Only the isolates confirmed as *E. coli* were further analyzed—774 isolates in total: 306 from WWTP Gdansk–Wschod (80 from WRW, 134 from WTW, and 92 from WMOut), 343 from WWTP Gdynia–Debogorze

(124 from DRW, 102 from DTW, and 117 from DMOut), and 125 from the mouth of the Vistula River (VR).

4.2.7 DNA EXTRACTION

Total DNA was obtained from several bacterial colonies, freshly grown on LB agar (tryptone 10 g/L, yeast extract 5 g/L, sodium chloride 10 g/L, agar 15 g/L; Becton, Dickinson and Company). Bacterial suspensions were prepared in distilled water and subsequently boiled in a water bath for 5 min. Cell debris was centrifuged at 13,000 rpm for 5 min, and the resulting supernatant was used as a template for all PCR assays.

4.2.8 IDENTIFICATION AND CHARACTERIZATION OF INTEGRONS

Isolates resistant to at least one antimicrobial agent (n = 262) were tested for the presence of integrons. These included 130 isolates from WWTP Gdansk–Wschod (35 from WRW, 49 from WTW, and 46 from WMOut), 103 from WWTP Gdynia–Debogorze (47 from DRW, 35 from DTW, and 21 from DMOut), and 29 from the mouth of the Vistula River (VR).

The presence of integrons was detected by PCR amplification of the integrase gene intI1 (for class 1 integrons) and intI2 (for class 2 integrons) using previously described primers (Kraft et al. 1986; Falbo et al. 1999; Mazel et al. 2000). Additionally, the presence of sul1 and qacEΔ1 genes, which are usually associated with integrons (Stokes and Hall 1989; Paulsen et al. 1993), was also checked. In order to determine the size of the variable gene cassette regions in integron-positive isolates, we used previously described primer sets 5CS–3CS and hepF–hepR (for class 1 and class 2 variable region, respectively) (Levesque et al. 1995; White et al. 2001). All primers used are shown in Table 1. All PCR reactions were carried out in T1 thermal cycler (Biometra GmbH, Goettingen, Germany) using nucleotides and buffers purchased from A&A Biotechnology. For detection of integrase genes, sul1 and qacEΔ1 genes Taq polymerase was used (A&A Biotechnology). The temperature profile for integrase, sul1, and qacEΔ1 genes amplification was as follows: initial denaturation (94 °C for 9 min), followed by 30 cycles of denaturation (94 °C for 30 s), annealing (30 s at temperature indicated in Table 1), and extension (72 °C for 1 min); and then a final extension (72 °C for 10 min). Selected class 1 and 2 integrase amplicons were sequenced and served as positive controls in further PCR experiments. Sequences of integrase genes from these positive controls were deposited in GenBank under accession numbers KM219981 and KM219982. *E. coli* DH5α strain served as negative control during PCR experiments.

In order to determine the size of variable regions of integrons, gene cassette amplifications were performed using Phusion™ polymerase (Thermo Fisher Scientific), and amplification was carried out as follows: initial denaturation (98 °C for 30 s), followed by 24 cycles of denaturation (98 °C for 10 s), annealing (63 °C for 30 s), elongation (72 °C for 6 min), and final elongation (72 °C for 10 min). PCR products were analyzed by electrophoresis on a 1 % agarose gel in $1 \times$ TAE buffer (Sambrook and Russell 2001) and stained with ethidium bromide (0.5 μg mL−1), visualized under UV light, and documented using Vilber Lourmat image acquisition system. Size of the PCR products was compared with 1-kb DNA Ladder (Thermo Scientific) in Bio1D software (Vilber Lourmat, Marnela-Vallée, France).

4.2.9 SEQUENCING OF GENE CASSETTE ARRAYS

Selected amplicons representing various size classes of integrons were sequenced using ABI PRISM BigDye Terminator cycle sequencing with forward and reverse primers (for class 1, 5CS–3CS, and for class 2 integrons, hepF–hepR) (Macrogen, Korea). Raw sequence reads were assembled using pregap4 and gap4 programs from Staden Package (Bonfield et al. 1995). High-quality consensus sequences were extracted from the assembly and afterward analyzed. DNA sequence analysis and gene cassette array annotations were performed using BLAST algorithm against INTEGRALL database (Moura et al. 2009). Sequences were deposited in GenBank under accession numbers KJ192400–KJ192434.

TABLE 1. Primers used in this study.

Target	Primer sequences (5'-3')	PCR product size (bp)	PCR annealing temperature (°C)	Reference
intI1	IntIAF: CCT CCC GCA CGATGATC IntIAR: TCC ACG CAT CGT CAG GC	280	55	Kraft et al. 1986
intI1	IntIBF: GAA GAC GGC TGC ACT GAA CG IntIBR: AAA ACC GCC ACT GCG CCG TTA	1,201	65	Falbo et al. 1999
intI2	Int2AF: TTATTG CTG GGATTA GGC Int2AR: ACG GCT ACC CTC TGT TAT C	233	50	Goldstein et al. 2001
intI2	Int2BF: GTA GCA AAC GAG TGA CGA AAT G Int2BR: CAC GGATAT GCG ACA AAA AGG T	788	65	Mazel et al. 2000
Class 1 integron variable region	5CS: GGC ATC CAA GCA GCA AG 3CS: AAG CAG ACT TGA CCT GA	Variable	63	Levesque et al. 1995
Class 2 integron variable region	hepF: CGG GAT CCC GGA CGG CAT GCA CGATTT GTA hepR: GAT GCC ATC GCA AGTACG AG	Variable	63	White et al. 2001
sul1	Sul1F: ATG GTG ACG GTG TTC GGC ATT CTG A Sul1R: CTA GGC ATG ATC TAA CCC TCG GTC T	800	55	Grape et al. 2003
qacEΔ1	qacF: ATC GCA ATA GTT GGC GAA GT qacR: CAA GCT TTT GCC CAT GAA GC	225	55	Stokes and Hall 1989

4.2.10 STATISTICAL ANALYSIS

Significance of differences between antibiotic-resistance rates in all studied samples was determined using Fisher's exact test. The frequencies of resistance to particular antimicrobials and presence of MDR phenotype in integron-positive and integron-negative isolates were compared with Fisher's exact test. The differences in antimicrobial-resistance ranges, expressed as the number of antimicrobials or antimicrobial classes, to which the isolates were resistant, were compared with Mann–Whitney U test. $P < 0.05$ was considered to indicate statistical significance. Calculations were performed with Statistica 7 software (StatSoft).

4.3 RESULTS AND DISCUSSION

4.3.1 E. COLI NUMBER

In this study, the impact posed by treated wastewater on the receiving waters was evaluated using *E. coli* isolates. The number of *E. coli* in raw wastewater (WRW, DRW) varied between 0.7×10^7 and 23×10^7 colony-forming unit (CFU) per 100 mL (Table 2). Both WWTPs demonstrated high efficiency in *E. coli* removal (over 99 %), although occasionally exceeding 10^5 CFU per 100 mL in effluents (WTW, DTW). In the case of their marine outfalls (WMOut, DMOut), as well as the mouth of the Vistula River (VR), only single *E. coli* cells (<100 CFU per 100 mL) were detected, indicating that the water quality was better than the "excellent" according to the New Bathing Water Directive 2006/7/EC.

4.3.2 E. COLI WITH ANTIBIOTIC AND MULTIPLE-ANTIBIOTIC RESISTANCE

One of the objectives of this study was to compare the antimicrobial susceptibility among *E. coli* of wastewater (WRW, WTW and DRW, DTW), marine water (WMOut and DMOut), and river mouth origin (VR). In total,

TABLE 2   Average number of *E. coli* (CFU 100 mL^{-1}) in raw (WRW, DRW) and treated wastewater (WTW, DTW) as well as in marine outfalls (WMOut, DMOut), and mouth of the Vistula River (VR) (minimal and maximal number of *E. coli* in brackets).

Gdansk–Wschod			Gdynia–Debogorze			Vistula River (VR)
WRW	WTW	WMOut	DRW	DTW	DMOut	
10.7 (0.7–23) $\times 10^7$	2.4 (0.1–6.1) $\times 10^5$	43 (30–60)	8.5 (0.7–13) $\times 10^7$	2.1 (0.8–3.1) $\times 10^5$	22 (2–32)	21 (3–25)

774 *E. coli* isolates were investigated. Regardless of the sampling point, high prevalence of AM-resistant *E. coli* was detected (Fig. 2). The resistance rate varied, however, in broad range, from 13 % in VR to 47 % in WMOut. The prevalence of penicillin-resistance among *E. coli* was also reported in clinical settings (ECDC 2011), as well as in different environmental compartments (Li et al. 2009). The resistance rate, however, varied significantly. In most European countries since 2007, the resistance to aminopenicillins has exceeded 40 % among *E. coli* of clinical origin (in Poland between 54 and 65 %) (ECDC 2011). The current state of knowledge suggests that the spread of antibiotic resistance is mainly due to selective pressure (ECDC/EMEA 2009; Martinez 2008; Baquero et al. 2009; Harada and Asai 2010; Andersson and Hughes 2012). Thus, these findings are not surprising—penicillins are the most often used antibacterial agents (ATC group J01) in the community (outside the hospital) in Europe (ECDC 2010). Also, broad-spectrum penicillins (J01CA) were the most consumed antibiotics in Poland—about five defined daily doses (DDD) per 1,000 inhabitants and per day, followed by combination of penicillins with β-lactamase inhibitors (J01CR)—about 3.5 DDDs per 1,000 inhabitants and per day in 2010 (ECDC 2010). In the present study, the prevalence of AM-resistant *E. coli*, followed by isolates resistant to combination of amoxicillin (penicillins) and clavulanate (β-lactamase inhibitor) (up to

32 %), seems to mirror the usage trends. Slightly lower resistance rates were noted to trimethoprim/sulfamethoxazole (5–20 %) and fluoroquinolones (5–15 %) (Fig. 2). Trimethoprim/sulfamethoxazole and fluoroquinolones are recommended as first-line drugs for uncomplicated urinary tract infections (UTIs). Recently, however, fluoroquinolones are increasingly being used instead of trimethoprim/sulfamethoxazole (Kallen et al. 2006). In consequence, selective pressure together with plasmid-mediated quinolone-resistance (PMQR) mechanisms has probably accelerated the rate of fluoroquinolone resistance spread among uropathogens (Gupta et al. 2005). In Europe, among clinical uropathogen *E. coli* strains, fluoroquinolones resistance exceeded 20 % (in Poland up to 26 % in 2010) (ECDC 2011). Quinolone and fluoroquinolone resistance is being increasingly reported around the world (Poirel et al. 2012). Principal mechanisms of resistance to those antibiotics are chromosome-encoded. However, the emergence of plasmid-mediated resistance is a major threat to public health (Mammeri et al. 2005; Poirel et al. 2012; Mokracka et al. 2012).

Among *E. coli* strains isolated from feces of mallards, herring gulls, and waterbirds on the Polish coast of the Baltic Sea, plasmid-encoded quinolone resistance associated with the qnrS gene was noted (Literak et al. 2010). Resistance to nalidixic acid was most frequent (together with resistance to tetracycline, followed by resistance to aminoglycosides and ampicillin) among heterotrophic marine bacteria isolated from Baltic Sea water samples (Moskot et al. 2012).

Rising trends are also observed among clinical *E. coli* resistant to third-generation cephalosporins and aminoglycosides. In this study, such resistance was reported only occasionally, while resistance to clinically relevant carbapenems (IPM and MEM) was not detected (Fig. 2). However, three isolates from raw wastewater (one from WRW and two from DRW) and two isolates from marine outfall (WMOut) produced ESBL. According to the ECDC surveillance report (2011) and numerous other reports (Pitout et al. 2005; Perez et al. 2007; Guenther et al. 2011; Liebana et al. 2013), the prevalence of ESBL-producing isolates has been on a continuous increase during the last decade. The majority of ESBLs are isolated from human clinical samples; however, they were also detected in wastewater and human-impacted environmental compartments (e.g., Galvin et al. 2010; Reinthaler et al. 2010; Hartmann et al. 2012;

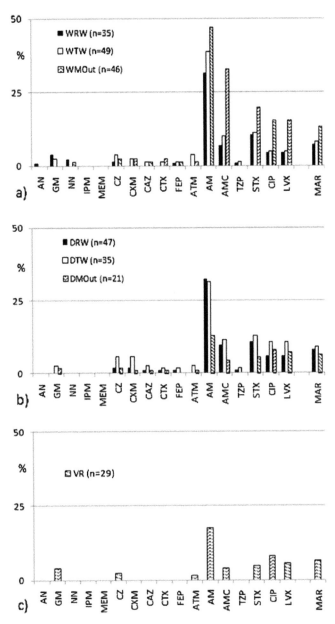

Figure 2. Antibiotic susceptibility of E. coli isolates from raw wastewater (WRW, DRW), treated wastewater (WTW, DTW), and marine outfalls (WMOut, DMOut) of a WWTP Gdansk–Wschod and b WWTP Gdynia–Debogorze, respectively, as well as from c the mouth of the Vistula River (VR).

Chagas et al. 2011; Korzeniewska et al. 2013). ESBL-producing isolates of domestic and wild animal origin were also reported (Smet et al. 2010; Guenther et al. 2011; Tausova et al. 2012; Liebana et al. 2013). *E. coli* isolates with MDR phenotype were detected in all samples, comprising from 6 % (in VR and DMOut) to 13 % (in WMOut) of isolates. The positive selection of bacteria with resistance patterns, previously described in wastewater processes (Łuczkiewicz et al. 2010; Ferreira da Silva et al. 2007; Novo and Manaia 2010), was also observed in this study for both of the WWTPs. Resistance rate to AMC, STX, CIP, and LVX noted for *E. coli* isolated from treated wastewater (WTW and DTW) was higher than that observed in corresponding raw wastewater (WRW and DRW), but the differences were not statistically significant (in Fisher's exact test). In the case of narrow- (CZ, CXM) and extended-spectrum (CAZ, CTX, FEP) cephalosporins, as well as TZP, the results were not evaluated due to the low number of resistant isolates. Interestingly, statistically significant rising trends (Fisher's exact test, $P < 0.05$) were noted for AMC-, STX-, CIP-, and LVX-resistant *E. coli*, as well as for *E. coli* with MDR pattern, isolated from marine outfall WMOut and corresponding treated wastewater (WTW). Such phenomenon was not observed for treated wastewater of WWTP Gdynia–Debogorze (DTW) and its marine outfall (DMOut). Resistance rates among *E. coli* from DMOut were comparable to those detected in the mouth of the Vistula River (VR). The differences between observed local impact caused by marine outfalls may be partly explained by differences in their operation time. Treated wastewater from WWTP Gdansk–Wschod was discharged via marine outfall (WMOut) for the last 12 years, while marine outfall of WWTP Gdynia–Debogorze (DMOut) has been operating for 2 years only. However, further detailed analyses are needed to understand better the impact of resistance genes in these water ecosystems.

Among 37 and 34 % of *E. coli* isolated from raw wastewater samples (WRW and DRW, respectively), resistance to at least one of the antimicrobial agents was noted (Fig. 3). In both WWTP effluents, the resistance rate increased to 44 % for WTW and 38 % for DTW. In the case of marine outfalls rising, trends were observed only for WMOut (47 % of *E. coli* isolates showed resistance), while resistance rate observed in DMOut was

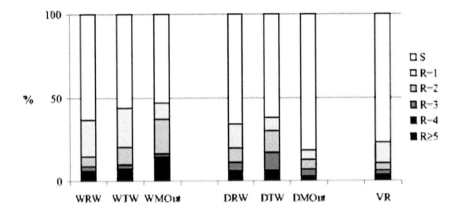

Figure 3. Susceptibility of *E. coli* isolates to increasing number of antimicrobial agents (S sensitive to all, R1-R≥5 resistant from 1 to 5 and more antimicrobial agents)

significantly lower (18 %) and similar to that noted for mouth of the VR (23 %).

As indicated in this study, human-associated bacteria with resistance and MDR phenotypes can survive wastewater treatment processes and be disseminated in the receiving waters. Since it is believed that WWTPs represent hotspots for horizontal gene transfer (Mokracka et al. 2012; Ma et al. 2013), we decided to assess the prevalence of class 1 and 2 integrons among antimicrobial-resistant *E. coli* isolated during this study.

4.3.3 CLASS 1 AND 2 INTEGRONS IN ANTIMICROBIAL-RESISTANT E. COLI

Since integrons have been associated with antibiotic-resistance capture and dissemination, in this study, the integron presence in *E. coli* isolates resistant to at least one antimicrobial agent was tested. In general, presence of class 1 and 2 integrons was detected in *E. coli* of 32.06 % (n = 84) and 3.05 % (n = 8), respectively. Both classes of integrons were detected in one isolate (0.38 %). Other authors reported smaller amounts of integron-positive isolates in wastewater environments, but in these reports, often all *Enterobacteriaceae* or *Aeromonadaceae* isolates were taken under consideration (e.g., Koczura et al. 2012; Moura et al. 2012). Nevertheless, similar frequencies of integron-bearing *E. coli* of animal, human, and wastewater origin were noted (Cocchi et al. 2007; Vinue et al. 2008). Xia et al. (2013) reported class 2 integrons in 8 % among antimicrobial-resistant gram-negative bacteria isolated from wastewater environments in China. In our study, we did not check clonal relatedness of the isolates; thus, it could bias the results concerning prevalence of integrons. However, similar (30.8 and 1 % for class 1 and 2, respectively) (Kang et al. 2005) or even higher (85.6 and 3.6 % for class 1 and 2, respectively) (Su et al. 2006) prevalence of integrons in *E. coli* from clinical samples was recorded. In those studies, only selected strains showing the same gene cassette patterns were checked for clonal relatedness using PFGE patterns or ERIC-PCR, revealing distinct patterns; thus, little clonal relatedness between studied strains was stated (Kang et al. 2005; Su et al. 2006).

In raw wastewater, WRW and DRW, class 1 integrons were detected in 28.6 and 38.3 % of resistant *E. coli*, respectively, whereas in treated wastewater, WTW and DTW, in 26.5 and 37.1 % of isolates, respectively. In the isolates collected from marine outfalls, WMOut and DMOut, class 1 integrons were present in 29 and 37.1 % of resistant isolates respectively, while in the mouth of the Vistula River (VR) in 27.6 % of isolates.

Class 2 integrons were significantly less frequent. They were detected only in raw wastewater obtained from WWTP Gdynia–Debogorze (DRW) (one isolate—2.1 %), as well as in the treated wastewater of WWTP Gdansk–Wschod (WTW) (four isolates—8.2 %). In marine outfalls of both WWTPs, two isolates with class 2 integron were isolated (one isolate from each of the sampling points). Also, in the mouth of the VR, class 2 integron was detected only in one isolate. Additionally, only in this particular isolate (isolate O5 5017), integrons from both classes were present together (Fig. 4, Table 2).

Among 84 isolates positive for intI1 gene, there were 16 isolates, in which only one pair of primers was successful in amplification of intI1

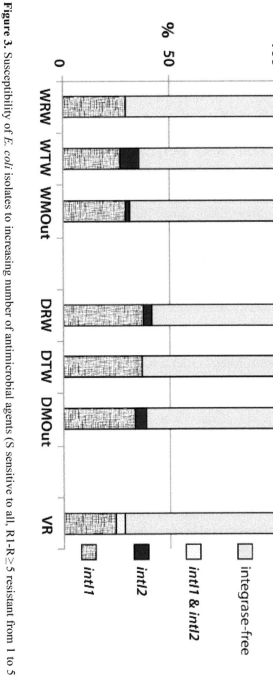

Figure 3. Susceptibility of *E. coli* isolates to increasing number of antimicrobial agents (S sensitive to all, R1-R≥5 resistant from 1 to 5 and more antimicrobial agents)

gene (with product size of 288 bp, Table 1). These isolates were nonetheless considered as intI1 positive because they gave positive results with primers designed to amplify variable region of class 1 integron. Among all class 2 integron-positive isolates (n = 8), intI2 gene was amplified with both primers designed for this integrase gene (Table 1).

4.3.4 ASSOCIATION BETWEEN ANTIBIOTIC RESISTANCE, MDR PHENOTYPE, AND INTEGRONS IN E. COLI ISOLATES

Among the integron-positive isolates, the highest resistance frequency was noted for ampicillin, amoxicillin/clavulanate, trimethoprim/sulfamethoxazole, and fluoroquinolones. Integron-negative isolates were also most frequently resistant to ampicillin, amoxicillin/clavulanate, and fluoroquinolones. Resistance to imipenem and meropenem was not observed in both groups of isolates, and none of the integron-negative isolates were resistant to amikacin. Resistance to fluoroquinolones (ciprofloxacin and levofloxacin), trimethoprim/sulfamethoxazole, amoxicillin/clavulanate, and piperacillin/tazobactam was significantly associated with the presence of integrons (Fisher's exact test, $P < 0.05$ for piperacillin/tazobactam, $P < 0.01$ for the other abovementioned antibiotics).

Presence of integrons together with STX resistance is not surprising, since occurrence of sulfonamide resistance gene (sul1) is often associated with classic class 1 integrons (Paulsen et al. 1993). Correlation of other resistance with integrons is likely to the presence of numerous resistance genes in one genetic element, e.g., plasmids or transposons (Stalder et al. 2012). Similar findings were observed in heterotrophic bacteria isolated from Baltic Sea waters. In those strains, statistical analysis indicated relationships between resistance to some antibiotics (ampicillin and erythromycin, chloramphenicol and erythromycin, chloramphenicol and tetracycline, erythromycin and tetracycline), suggesting the linkage of resistance genes for antibiotics belonging to different classes (Moskot et al. 2012). No statistically significant differences were found for cephalosporins, carbapenems, aminoglycosides, and aztreonam, due to lack of or a small number of isolates resistant to those antibiotics in the study. The integron-bearing isolates were resistant to 1–13 of the tested antimicrobials belonging to 1–8 classes, and 50 % of those isolates were multidrug resistant (according to the definition proposed by Magiorakos et al. (2012)). Integron-negative isolates were resistant to 1–11 antimicrobials belonging to 1–7 classes and 17 of them (10 %) were multidrug resistant. The difference in the MDR pheno-type frequency between isolates with and without integrons was statistically significant (P<0.01, Fisher's exact test). This is a common phenomenon, in-dependent of species or origin of samples (Leverstein-van Hall et al. 2003). The differences in resistance ranges (number of antimicrobials or number of antimicrobial classes) between both groups of isolates were statistically significant (P<0.001, Mann–Whitney U test).

4.3.5 CHARACTERISTICS OF CLASS 1 AND 2 VARIABLE REGIONS AND GENE CASSETTE ARRAYS

E. coli isolates carrying the class 1 and 2 integrons (84 and 8 isolates, respectively) were further analyzed for the presence of inserted gene cas-settes in the variable region using the primer sets 5CS–3CS and hepF–hepR (Table 1), designed to amplify variable region of integrons class 1 and 2, respectively. Among all class 1 integron-positive isolates, 57.14 % (n=48) possessed variable regions with size ranging from 650 to 2,600 bp. Among eight isolates bearing class 2 integrons, five isolates (62.5 %) had variable region with size ranging from 2,200 to 2,500 bp. Characteristics of those integron-bearing isolates are shown in Table 3.

In some integrase-positive isolates, variable regions were not detected (with standard primers 5CS and 3CS). The absence of gene cassettes in those isolates could be possible due to the presence of a cassette array that is too large to amplify (Partridge et al. 2009). Similar findings were also noted by Ndi and Barton (2011) and Yu et al. (2003). Since occur-rence of sulfonamide resistance gene (sul1) and quaternary ammonium compounds resistance gene (qacEΔ1) is often associated with classic class 1 integrons (Paulsen et al. 1993), in all intI1-positive isolates, presence of those genes was tested. In 25 isolates (29.76 %), sul1 gene was not de-tected, and in 23 of those isolates, we failed to amplify the variable regions.

TABLE 3. Characteristics of integron-bearing *E. coli* isolates isolated from WWTP "Gdansk–Wschod" and WWTP "Gdynia–Debogorze" (raw wastewater: WRW and DRW; treated wastewater: WTW and DTW), and WWTPs marine outfalls (WMOut and DMOut) and also from mouth of the Vistula River (VR)

Isolate no.	Class of integron	sul1	qacEΔ1	Antimicrobial resistance tested		Integron variable regions	
				No. of resistances	Resistance phenotype	Size of amplicon (bp)	Gene cassette array
WRW (raw wastewater from "Gdansk–Wschod" WWTP)							
GCK 27	1	+	+	2	AM STX	1,600	dfrA17-aadA5
GCK 140	1	+	+	1	GM	1,500	dfrA1-aadA1
GCK 141	1	+	+	2	AM STX	1,600	dfrA1-aadA1
GCK 149	1	+	+	3 (MDR)	CZ CXM AM STX	1,650	NT
GCR 150	1	+	+	3 (MDR)	AM AMC STX	1,550	NT
WTW (treated wastewater from "Gdansk–Wschod" WWTP)							
WCO 102	2	-	-	3 (MDR)	AM AMC STX	2,200	dfrA1-sat2-aadA1[a]
GCO 54	2	-	-	3 (MDR)	AM AMC STX	2,300	dfrA1-sat2-aadA1[a]
GCO 55	1	+	+	1	AM	1,000	aadA1
GCO 181	1	+	+	1	STX	1,500	dfrA1-aadA1
GCO 183	1	+	+	3 (MDR)	AM STX CIP LVX	1,500	dfrA1-aadA1
WCUV 522	1	+	+	3 (MDR)	AM STX GM NN	1,550	dfrA17-aadA5
GCO 175	1	+	+	2	AM STX	1,450	NT

TABLE 3. CONTINUED.

WCO 3	1	+	+	1	AM	950	NT
GCO 164	1	+	+	4 (MDR)	AM STX CIP LVX GM	1,550	NT
GCO 187	1	+	+	1	CIP LVX	1,500	NT
GCO 188	1	+	+	1	AM	1,600	NT
WCUVO 502	1	+	+	1	AM	1,000	NT
WCUV 523	1	+	+	2	AM STX	1,500	NT
WMOut (marine outfall of "Gdansk–Wschod" WWTP)							
WWDC 727	2	–	–	6 (MDR, ESBL,)	CZ CXM CAZ CTX FEP ATM AM AMC STX	2,500	dfrA1-sat2-aadA1[a]
WWDC 719	1	+	+	4 (MDR)	AM AMC STX CIP LVX	1,500	dfrA1-aadA1
WWDC 701	1	–	+	2	AM STX	650	dfrA7
WWDC 743	1	+	+	2	AM STX	1,550	dfrA17-aadA5
WWDC 755A	1	+	+	4 (MDR)	AM AMC STX CIP LVX	1,450	NT
WWDC 755B	1	+	+	4 (MDR)	AM AMC STX CIP LVX	1,450	NT
WWDC 762A	1	+	+	4 (MDR)	AM AMC STX CIP LVX	2,000	NT
WWDC 762B	1	+	+	4 (MDR)	AM AMC STX CIP LVX	2,000	NT
WWDC 763	1	+	+	4 (MDR)	AM AMC STX CIP LVX	1,700	NT
WWDC 817	1	+	+	2	STX CIP LVX	2,000	NT
WWDC 829	1	+	+	7 (MDR, ESBL,)	CZ CXM CAZ CTX FEP AM AMC STX CIP LVX NN	1,800	NT

TABLE 3. CONTINUED.

WWDC 845A	1	+	+	4 (MDR)	AM AMC STX CIP LVX	1,700	NT
DRW (raw wastewater from "Gdynia–Debogorze" WWTP)							
BMK 231	1	+	+	3 (MDR)	AM STX CIP LVX	1,550	dfrA1-aadA1
BMK 236	1	+	+	2	AM STX	1,600	dfrA17-aadA5
BMS 512	1	+	+	3 (MDR)	AM AMC STX	1,900	bla$_{OXA30}$-aadA1
BMS 606	1	+	+	3 (MDR)	AM STX CIP LVX	1,550	dfrA17-aadA5
BMK 639	1	+	+	2	AM STX	1,800	dfrA12-orfF-aadA2
BMS 710	1	+		7 (MDR, ESBL)	CZ CXM CAZ CTX FEP ATM AM AMC STX CIP LVX	1,500	dfrA1-aadA1
BMS 719	1	+	+	4 (MDR)	AM AMC STX GM	1,600	dfrA17-aadA5
BMK 834	1	+	+	2	CIP LVX GM	1,000	aadA1
BMS 2009	1	+	+	2	AM STX	1,900	dfrA12-orfF-aadA2
BMK 2011	1	+	+	3 (MDR)	AM AMC STX	1,900	dfrA12-orfF-aadA2
BMS 4003	1	+	+	2	AM STX	1,500	dfrA1-aadA1
BMK 4016	1	+	+	2	AM STX	1,500	dfrA1-aadA1
BMK 26	1	+	+	3 (MDR)	AM AMC TZP	1,950	NT
DTW (treated wastewater from "Gdynia–Debogorze" WWTP)							

TABLE 3. CONTINUED.

BMO 244	1	+	4 (MDR)	AM AMC STX TZP	1,500	dfrA1-aadA1
BMO 645	1	+	2	AM STX	700	dfrA5
BMO 843	1	+	1	AM	1,000	aadA1
BMO 3023	1	+	2	AM STX	1,550	dfrA1-aadA1
BMO 4023	1	+	1	AM	1,000	aadA1
BMO 241	1	+	3 (MDR)	AM STX CIP LVX	2,500	NT
BMO 5022	1	+	3 (MDR)	AM STX CIP LVX	2,000	NT
DCUV 14	1	+	2	AM STX	1,600	NT
DCUV 105	1	+	3 (MDR)	AM AMC STX	1,000	NT
DMOut (marine outfall of "Gdynia–Debogorze" WWTP)						
R1 3110	1	+	2	AM STX	1,500	dfrA1-aadA1
MWDC 2202	2	–	3 (MDR)	AM AMC STX	2,500	NT
MWDC 2203	1	–	1	AM	800	NT
R3 5031	1	+	3 (MDR)	AM STX CIP LVX	1,400	NT
VR (mouth of the Vistula River)						
O4 2092	1	+	1	AM	1,000	aadA1
O4 5012	1	+	1	AM	1,000	aadA1
O5 5017	1	+	2	AM STX	1,500	dfrA1-aadA1
	2				2,600	estX-sat2[a]

TABLE 3. CONTINUED.

UWWC 3001A	1	+	+	4 (MDR)	AM AMC STX CIP LVX	1,700	NT
UWWC 3001B	1	+	+	4 (MDR)	AM AMC STX CIP LVX	1,750	NT
O3 5008	1	+	+	3 (MDR)	AM AMC STX	1,500	NT

AM ampicillin, AMC amoxicillin/clavulanate, ATM aztreonam, CZ cefazolin, CAZ ceftazidime, CTX cefotaxime, CXM cefuroxime, FEP cefepime, CIP ciprofloxacin, LVX levofloxacin, GM gentamicin, NN tobramycin, TZP piperacillin/tazobactam, STX trimethoprim/sulfamethoxazole, NT not tested, MDR multidrug-resistance phenotype, ESBL extended-spectrum β-lactamase phenotype

[a]Only partial sequence was available

In 21 isolates (25 %), qacEΔ1 gene was not detected, and those isolates also did not possess variable regions. Similar findings were noted for P. aeruginosa (Nass et al. 1998) and *E. coli* (Sáenz et al. 2004). However, there were two isolates (WWDC 701 from WMOut and MWDC 2203 from DMOut, Table 2) in which we managed to amplify variable regions, despite the lack of sul1 gene and the presence of qacEΔ1 gene. There were also three isolates with mentioned above 3′ end of the integron, in which variable regions were not detected. Thus, we assume that in isolates without detectable variable regions, there are integrons with non-classic structure, probably with the tni region or various insertion sequences (IS), as mentioned by Partridge et al. (2009).

Among all integron-positive isolates, 48 isolates with class 1 integron and 5 isolates with class 2 integron contained variable regions (listed in Table 3), but only 35 selected amplicons were sequenced. Further analysis of the obtained sequences showed that the selected isolates harbored nine distinct gene cassette arrays (Table 4).

The most prevalent genes detected in variable regions of integrons were those connected with resistance to aminoglycosides (aadA1, aadA2, and aadA5) and trimetophrim (dfrA1, dfrA5, dfrA7, and dfrA17). Those genes were present alone, as well as in combination with each other, or other resistance genes. Genes conferring resistance to streptothricin (estX and sat2) were detected only in isolates bearing class 2 integron. One isolate, obtained from the mouth of the VR, harbored both classes of integrons, with gene cassette arrays: dfrA1-aadA1 (class 1 integron) and estX-sat2 (class 2 integron). Similar patterns were detected in Aeromonas and Enterobacteriaceae strains isolated from slaughterhouse WWTP (Moura et al. 2007), in *E. coli* of human and animal origin in Korea (Kang et al. 2005) and *E. coli* from Seine Estuary in France (Laroche et al. 2009). Among all sequenced variable regions of integrons, the most prevalent cassette arrays were dfrA1-aadA1, dfrA17-aadA5, and aadA1 (Table 4).

It is interesting to find that some patterns were found only in isolates from raw wastewater, like dfrA12-orfF-aadA2 while others only in treated wastewater or marine outfalls of WWTPs (dfrA5 and dfrA7, respectively) (Table 4).

TABLE 4. Distribution of the different cassette arrays in E. coli isolates obtained from different sources.

Type of integron and gene cassette arrays		No. of isolates	Source of isolates			
Class 1	Class 2		RW	TW	MOut	VR
dfrA1-aadA1	–	13	6	4	2	1
dfrA17-aadA5	–	6	4	1	1	–
aadA1	–	6	2	3	–	1
dfrA12-orfF-aadA2	–	3	3	–	–	–
–	dfrA1-sat2-aadA1	3	–	2	1	–
dfrA7	–	1	–	–	1	–
dfrA5	–	1	–	1	–	–
bla$_{OXA30}$-aadA1	–	1	1	–	–	–
dfrA1-aadA1	estX-sat2[a]	1	–	–	–	1

RW raw wastewater, TW treated wastewater, MOut marine outfall of WWTPs, VR mouth of the Vistula River
[a]Only partial sequence was available

4.4 CONCLUSIONS

Data obtained in this study indicated that wastewater treatment processes together with effective dilution of treated wastewater by marine outfall were generally sufficient to protect coastal water quality from sanitary degradation. However, human-associated bacteria, even potential pathogens and bacteria carrying antibiotic resistance genes of clinical significance, survived in wastewater and marine water conditions. Moreover, these resistant bacteria were enriched in highly diverse integrons, with nine different gene cassette arrays. Statistically significant association between resistance to fluoroquinolones, trimethoprim/sulfamethoxazole, amoxicillin/clavulanate, and piperacillin/tazobactam and the presence of integrons in *E. coli* isolates was noted. Integrons are clearly important for the development of the MDR phenotype. In conclusion, data obtained during this study indicate the potential of WWTP's effluents in facilitating horizontal gene transfer of mobile genetic elements and MDR phenotypes in the studied area. Given the similarity of WWTP processes and municipal wastewater composition, it is likely the common problem requiring further investigation.

REFERENCES

1. Andersson DI, Hughes D (2012) Evolution of antibiotic resistance at non-lethal drug concentrations. Drug Resist Update 15:162–172
2. APHA (1998) Standard methods for the examination of water and wastewater, 20th edn. American Public Health Association, Washington, DC
3. Baquero F, Alvarez-Ortego C, Martinez JL (2009) Ecology and evolution of antibiotic resistance. Environ Microbiol Rep 1:469–476
4. Bonfield JK, Smith KF, Staden R (1995) A new DNA sequence assembly program. Nucleic Acids Res 23:4992–4999
5. Boucher Y, Labbate M, Koenig JE, Stokes HW (2007) Integrons: mobilizable platforms that promote genetic diversity in bacteria. Trends Microbiol 15:301–309
6. Cambray G, Guerout A-M, Mazel D (2010) Integrons. Annu Rev Genet 44:141–166
7. Chagas TPG, Seki LM, Cury JC, Oliveira JAL, Davila AMR, Silva DM, Asensi MD (2011) Multiresistance, beta-lactamase-encoding genes and bacterial diversity in hospital wastewater in Rio de Janeiro, Brazil. J Appl Microbiol 111:572–581
8. Cocchi S, Grasselli E, Gutacker M, Benagli C, Convert M, Piffaretti J-C (2007) Distribution and characterization of integrons in Escherichia coli strains of animal and human origin. FEMS Immunol Med Microbiol 50:126–132
9. Czekalski N, Berthold T, Caucci S, Egli A, Bürgmann H (2012) Increased levels of multiresistant bacteria and resistance genes after wastewater treatment and their dissemination into Lake Geneva, Switzerland. Front Microbiol. doi:10.3389/fmicb.2012.00106
10. D'Costa VM, McGrann KM, Hughes DW, Wright GD (2006) Sampling the antibiotic resistome. Science 311:374–377
11. DIRECTIVE 2006/7/EC of the European Parliament and of the Council (2006) Concerning the management of bathing water quality and repealing Directive 76/160/EEC.   http://eur-lex.europa.eu/LexUriServ/LexUriServ.do?uri=OJ:L:2006:064:0037:0051:EN:PDF. Accessed July 2014
12. Dubois V, Parizano M-P, Arpin C, Coulange L, Bezian MC, Quentin C (2007) High genetic stability of integrons in clinical isolates of Shigella spp. of worldwide origin. Antimicrob Agents Chemother 51:1333–1340

13. ECDC (2010) Survivallence of antimicrobial consumption in Europe. http://www.ecdc.europa.eu/en/publications/Publications/antimicrobial-antibiotic-consumption-ESAC-report-2010-data.pdf. Accessed July 2014

14. ECDC (2011) Antimicrobial resistance surveillance in Europe 2011. http://ecdc.europa.eu/en/publications/Publications/Forms/ECDC_DispForm.aspx?ID=998. Accessed July 2014

15. ECDC/EMEA (2009) Joint technical report: the bacterial challenge: time to react. http://www.ecdc.europa.eu/en/publications. Accessed July 2014

16. Elsaied H, Stokes HW, Nakamura T, Kitamura K, Kamagata Y, Fuse H, Maruyama A (2007) Novel and diverse integron integrase genes and integron-like gene cassettes are prevalent in deep-sea hydrothermal vents. Environ Microbiol 9:2298–2312

17. Elsaied H, Stokes HW, Kitamura K, Kurusu Y, Kamagata Y, Maruyama A (2011) Marine integrons containing novel integrase genes, attachment sites, attI, and associated gene cassettes in polluted sediments from Suez and Tokyo Bays. ISME J 5:1162–1177

18. EUCAST (2011) The European Committee on Antimicrobial Susceptibility Testing. http://www.eucast.org/clinical_breakpoints/. Accessed July 2014

19. Falbo V, Carattoli A, Tosini F, Pezzella C, Dionisi AM, Luzzi I (1999) Antibiotic resistance conferred by a conjugative plasmid and a class I integron in Vibrio cholerae O1 El Tor strains isolated in Albania and Italy. Antimicrob Agents Chemother 43:693–696

20. Ferreira da Silva M, Vaz-Moreira I, Gonzalez-Pajuelo M, Nunes OC, Manaia CM (2007) Antimicrobial resistance patterns in Enterobacteriaceae isolated from an urban wastewater treatment plant. FEMS Microbiol Ecol 60:166–176

21. Galvin S, Boyle F, Hickey P, Vellinga A, Morris D, Cormican M (2010) Enumeration and characterization of antimicrobial-resistant Escherichia coli bacteria in effluent from municipal, hospital, and secondary treatment facility sources. Appl Environ Microbiol 76:4772–4779

22. Gillings MR, Holley MP, Stokes HW (2009) Evidence for dynamic exchange of qac gene cassettes between class1integrons and other integrons in freshwater biofilms. FEMS Microbiol Lett 296:282–288

23. Goldstein C, Lee MD, Sanchez S, Hudson C, Philips B, Register B, Grady M, Liebert C, Summers AO, White DG, Maurer JJ (2001) Incidence of class 1 and 2 integrases in clinical and commensal bacteria from livestock, companion animals, and exotics. Antimicrob Agents Chemother 45:723–726

24. Grape M, Sundström L, Kronvall G (2003) Sulphonamide resistance gene sul3 found in Escherichia coli isolates from human sources. J Antimicrob Ther 52:1022–1024

25. Guenther S, Ewers C, Wieler LH (2011) Extended-spectrum beta-lactamases producing E. coli in wildlife, yet another form of environmental pollution? Front Microbiol. doi:10.3389/fmicb.2011.00246

26. Gupta K, Hooton TM, Stamm WE (2005) Isolation of fluoroquinolone-resistant rectal Escherichia coli after treatment of acute uncomplicated cystitis. J Antimicrob Chemother 56:243–246

27. Harada K, Asai T (2010) Role of antimicrobial selective pressure and secondary factors on antimicrobial resistance prevalence in Escherichia coli from food-producing animals in Japan. J Biomed Biotechnol. doi:10.1155/2010/180682

28. Hartmann A, Locatelli A, Amoureux L, Depret G, Jolivet C, Gueneau E, Neuwirth C (2012) Occurrence of CTX-M producing Escherichia coli in soils, cattle, and farm environment in France (Burgundy Region). Front Microbiol. doi:10.3389/fmicb. 2012.00083

29. Iwane T, Urase T, Yamamoto K (2001) Possible impact of treated wastewater discharge on incidence of antibiotic resistant bacteria in river water. Water Sci Technol 43:91–99

30. Kallen AJ, Welch HG, Sirovich BE (2006) Current antibiotic therapy for isolated urinary tract infections in women. Arch Intern Med 166:635–639

31. Kang HY, Jeong YS, Oh JY, Tae SH, Choi CH, Moon DC, Lee WK, Lee YC, Seol SY, Cho DT, Lee JC (2005) Characterization of antimicrobial resistance and class 1 integrons found in Escherichia coli isolates from humans and animals in Korea. J Antimicrob Chemother 55:639–644

32. Koczura R, Mokracka J, Jablonska L, Gozdecka E, Kubek M, Kaznowski A (2012) Antimicrobial resistance of integron-harboring Escherichia coli isolates from clinical samples, wastewater treatment plant and river water. Sci Total Environ 414:680–685

33. Korzeniewska E, Korzeniewska A, Harnisz M (2013) Antibiotic resistant Escherichia coli in hospital and municipal sewage and their emission to the environment. Ecotox Environ Safe 91:96–102

34. Kraft CA, Timbury MC, Platt DJ (1986) Distribution and genetic location of Tn7 in trimethoprim-resistant Escherichia coli. J Med Microbiol 22:25–131

35. Laroche E, Pawlak B, Berthe T, Skurnik D, Petit F (2009) Occurrence of antibiotic resistance and class 1, 2 and 3 integrons in Escherichia coli isolated from a densely populated estuary (Seine, France). FEMS Microbiol Ecol 68:118–130

36. Leverstein-van Hall MA, Blok HEM, Donders ART, Paauw A, Fluit AC, Verhoef J (2003) Multidrug resistance among Enterobacteriaceae is strongly associated with the presence of integrons and is independent of species or isolate origin. J Infect Dis 187:251–259

37. Levesque C, Piche L, Larose C, Roy PH (1995) PCR mapping of integrons reveals several novel combinations of resistance genes. Antimicrob Agents Chemother 39:185–191

38. Li D, Yu T, Zhang Y, Yang M, Li Z, Liu M, Qi R (2009) Antibiotic resistance characteristics of environmental bacteria from an oxytetracycline production wastewater treatment plant and the receiving river. Appl Environ Microbiol 76:3444–3451

39. Liebana E, Carattoli A, Coque TM, Hasman H, Magiorakos AP, Mevius D, Peixe L, Poirel L, Schuepbach-Regula G, Torneke K, Torren-Edo J, Torres C, Threlfall J (2013) Public health risks of enterobacterial isolates producing extended-spectrum β-lactamases or AmpC β-lactamases in food and food-producing animals: an EU perspective of epidemiology, analytical methods, risk factors, and control options. Clin Infect Dis 56:1030–1037

40. Literak I, Dolejska M, Janoszowska D, Hrusakova J, Meissner W, Rzyska H, Bzoma S, Cizek A (2010) Antibiotic-resistant Escherichia coli bacteria, including strains with genes encoding the extended-spectrum beta-lactamase and QnrS, in waterbirds on the Baltic Sea coast of Poland. Appl Environ Microbiol 76:8126–8134

41. Łuczkiewicz A, Jankowska K, Fudala-Książek S, Olańczuk-Neyman K (2010) Antimicrobial resistance of fecal indicators in municipal wastewater treatment plant. Water Res 44:5089–5097

42. Ma L, Zhang X-X, Zhao F, Wu B, Cheng S, Yang L (2013) Sewage treatment plant serves as a hot-spot reservoir of integrons and gene cassettes. J Environ Biol 34:391–399

43. Magiorakos A-P, Srinivasan A, Carey RB, Carmeli Y, Falagas ME, Giske CG, Harbarth S, Hindler JF, Kahlmeter G, Olsson-Liljequist B, Paterson DL, Rice LB, Stelling J, Struelens MJ, Vatopoulos A, Weber JT, Monnet DL (2012) Multidrug resistant, extensively drug-resistant and pandrug-resistant bacteria: an international expert proposal for interim standard definitions for acquired resistance. Clin Microbiol Infect 18:268–281

44. Mammeri H, Van De Loo M, Poirel L, Martinez-Martinez L, Nordmann P (2005) Emergence of plasmid-mediated quinolone resistance in Escherichia coli in Europe. Antimicrob Agents Chemother 49:71–76

45. Martinez JL (2008) Antibiotics and antibiotic resistance genes in natural environments. Science 321:365–367

46. Mazel D, Dychinco B, Webb VA, Davies J (2000) Antibiotic resistance in the ECOR collection: integrons and identification of a novel aad gene. Antimicrob Agents Chemother 44:1568–1574

47. Mokracka J, Koczura R, Kaznowski A (2012) Multiresistant Enterobacteriaceae with class 1 and class 2 integrons in a municipal wastewater treatment plant. Water Res 46:3353–3363

48. Moskot M, Kotlarska E, Jakóbkiewicz-Banecka J, Gabig-Cimińska M, Fari K, Węgrzyn G, Wróbel B (2012) Metal and antibiotic resistance of bacteria isolated from the Baltic Sea. Int Microbiol 15:131–139

49. Moura A, Henriques I, Rieiro R, Correia A (2007) Prevalence and characterization of integrons from bacteria isolated from a slaughterhouse wastewater treatment plant. J Antimicrob Chemother 60:1243–1250

50. Moura A, Soares M, Pereira C, Leitão N, Henriques I, Correia A (2009) INTE-GRALL: a database and search engine for integrons, integrases and gene cassettes. Bioinformatics 25:1096–1098

51. Moura A, Pereira C, Henriques I, Correia A (2012) Novel gene cassettes and integrons in antibiotic-resistant bacteria isolated from urban wastewaters. Res Microbiol 163:92–100

52. Nass T, Sougakof W, Casetta A, Nordmann P (1998) Molecular characterization of OXA-20, a novel class D β-lactamase, and its integron from Pseudomonas aeruginosa. Antimicrob Agents Chemother 42:2074–2083

53. Ndi OL, Barton MD (2011) Incidence of class 1 integron and other antibiotic resistance determinants in Aeromonas spp. from rainbow trout farms in Australia. J Fish Dis 34:589–599

54. Novo A, Manaia CM (2010) Factors influencing antibiotic resistance burden in municipal wastewater treatment plants. Appl Microbiol Biotechnol 87:1157–1166

55. Pan J-C, Ye R, Meng D-M, Zhang W, Wang H-Q, Liu K-Z (2006) Molecular characteristics of class 1 and 2 integrons and their relationship to antibiotic resistance in

clinical isolates of Shigella sonnei and Shigella flexneri. J Antimicrob Chemother 58:288–296

56. Partridge SR, Tsafnat G, Coiera E, Iredell JR (2009) Gene cassette and cassette arrays in mobile resistance integrons. FEMS Microbiol Rev 33:757–784

57. Paulsen IT, Littlejohn TG, Rådström P, Sundström L, Sköld O, Swedberg G, Skurray RA (1993) The 3′ conserved segment of integrons contains a gene associated with multidrug resistance to antiseptics and disinfectants. Antimicrob Agents Chemother 37:761–768

58. Perez F, Endimiani A, Hujer KM, Bonomo RA (2007) The continuing challenge of ESBLs. Curr Opin Pharmacol 7:459–469

59. Pitout JDD, Nordmann P, Laupland KB, Poirel L (2005) Emergence of Enterobacteriaceae producing extended-spectrum β-lactamases (ESBLs) in the community. J Antimicrob Chemother 56:52–59

60. Poirel L, Cattoir V, Nordmann P (2012) Plasmid-mediated quinolone resistance; interactions between human, animal, and environmental ecologies. Front Microbiol. doi:10.3389/fmicb.2012/00024

61. Recchia GD, Hall RM (1995) Gene cassettes: a new class of mobile element. Microbiology 141:3015–3027

62. Reinthaler FF, Posch J, Feierl G, Wust G, Haas D, Ruckenbauer G, Mascher F, Marth E (2003) Antibiotic resistance of E. coli in sewage and sludge. Water Res 37:1685–1690

63. Reinthaler FF, Feierl G, Galler H, Haas D, Leitner E, Mascher F, Melkes A, Posch J, Winter I, Zarfel G, Marth E (2010) ESBL-producing E. coli in Austrian sewage sludge. Water Res 44:1981–1985

64. Rosser SJ, Young H-K (1999) Identification and characterization of class 1 integrons in bacteria from aquatic environment. J Antimicrob Chemother 44:11–18

65. Sáenz Y, Briñas L, Domínguez E, Ruiz J, Zarazaga M, Vila J, Torres C (2004) Mechanisms of resistance in multiple-antibiotic resistant Escherichia coli strains of human, animal, and food origins. Antimicrob Agents Chemother 48:3996–4001

66. Sambrook J, Russell DW (2001) Molecular cloning: a laboratory manual, 3rd edn. Cold Spring Harbour Laboratory Press, Cold Spring Harbour, New York

67. Smet A, Martel A, Persoons D, Dewulf J, Heyndrickx M, Herman L, Haesebrouck F, Butaye P (2010) Broad-spectrum beta-lactamases among Enterobacteriaceae of animal origin: molecular aspects, mobility and impact on public health. FEMS Microbiol Rev 34:95–316

68. Stalder T, Barraud O, Casellas M, Dagot C, Ploy M-C (2012) Integron involvement in environmental spread of antibiotic resistance. Front Microbiol. doi:10.3389/fmicb.2012.00119

69. Stokes HW, Hall RM (1989) A novel family of potentially mobile DNA elements encoding site-specific gene-integration functions: integrons. Mol Microbiol 3:1669–1683

70. Su J, Shi L, Yang L, Xiao Z, Li X, Yamasaki S (2006) Analysis of integrons in clinical isolates of Escherichia coli in China during the last six years. FEMS Microbiol Lett 254:75–80

71. Tausova D, Dolejska M, Cizek A, Hanusova L, Hrusakova J, Svoboda A, Camlik G, Literak I (2012) Escherichia coli with extended-spectrum β-lactamase and

plasmid-mediated quinolone resistance genes in great cormorants and mallards in Central Europe. J Antimicrob Chemother 67:1103–1107

72. Tchobanoglous G, Burton FL, Metcalf & Eddy (1991) Wastewater engineering: treatment, disposal, and reuse, 3rd edn. McGraw-Hill, New York

73. Vinue L, Saenz Y, Somalo S, Escudero E, Moreno MA, Ruiz-Larrea F, Torres C (2008) Prevalence and diversity of integrons and associated resistance genes in faecal Escherichia coli isolates of healthy humans in Spain. J Antimicrob Chemother 62:934–937

74. White PA, MacIver CJ, Rawlinson WD (2001) Integrons and gene cassettes in the Enterobacteriaceae. Antimicrob Agents Chemother 45:2658–2661

75. Xia R, Ren Y, Guo X, Xu H (2013) Molecular diversity of class 2 integrons in antibiotic-resistant gram-negative bacteria found in wastewater environments in China. Ecotoxicology 22:402–414

76. Xu Z, Li L, Shi L, Shirtliff ME (2001) Class 1 integron in staphylococci. Mol Biol Rep 38:5261–5279

77. Xu Z, Li L, Shirtliff ME, Peters BM, Peng Y, Alam MJ, Yamasaki S, Shi L (2010) First report of class 2 integron in clinical Enterococcus faecalis and class 1 integron in Enterococcus faecium in South China. Diagn Microbiol Infect Dis 68:315–317

78. Yu HS, Lee JC, Kang HY, Ro DW, Chung JY, Jeong YS, Tae SH, Choi CH, Lee EY, Seol SY, Lee YC, Cho DT (2003) Changes in gene cassettes of class 1 integrons among Escherichia coli isolates from urine specimens collected in Korea during the last two decades. J Clin Microbiol 41:5429–5433

CHAPTER 5

Prevalence of Antibiotic Resistance Genes and Bacterial Community Composition in a River Influenced by a Wastewater Treatment Plant

ELISABET MARTI, JUAN JOFRE, AND JOSE LUIS BALCAZAR

5.1 INTRODUCTION

Antibiotic resistance represents a significant global health problem due to the use and misuse of antibiotics, which favors the emergence and spread of resistant bacteria. Since the first warning of antibiotic resistance [1], this phenomenon has increased dramatically and as a result, 70% of all hospital-acquired infections in the United States are resistant to at least one family of antibiotics [2]. The treatment of these infections leads to higher healthcare costs because these therapies require longer hospital stays and more expensive drugs. To confront this increasing problem, it is necessary to understand the ecology of antibiotic resistance, including their origins, evolution, selection and dissemination [3].

Although antibiotic resistance has involved extensive research in clinically relevant human pathogens, environmental reservoirs of antibiotic resistance determinants and their contribution to resistance in clinical settings have only been considered in the last decade [4–6]. It has been shown that antibiotic resistance genes (ARGs) have environmental origins but the introduction and accumulation of antimicrobials in the environment facilitates their spread [7]. As a consequence, ARGs can be found in almost all environments and they are currently considered as emerging pollutants [8,9]. Therefore, identifying sources of resistance genes, their environmental distribution and how anthropogenic inputs affect their spread will aid in establishing strategies to combat antibiotic resistance.

The location of ARGs on genetic elements that can be mobilized, such as transposons, integrons and plasmids, facilitates the transfer of resistance to other organisms of the same or different species [10]. Although antibiotic resistance studies have been focused on cultivable bacteria and/ or indicator organisms in treated wastewater, the vast majority of environmental bacteria cannot be cultured under standard laboratory conditions. As a result, there is little information about how the discharge of wastewater effluents can affect bacterial communities and impact the prevalence of resistance genes in the environment.

It is well known that WWTPs reduce the total number of bacteria, especially coliforms, in their final effluent [11]. However, the treatment is not efficient enough to remove ARGs that are released to the receiving river [12,13]. In addition, WWTPs link human activities and the environment and may facilitate horizontal transfer of resistance determinants among a rich diversity of commensals, environmental microorganisms and clinically relevant pathogens [14]. In this regard, WWTP may contribute to the occurrence, spread and persistence of both antibiotic-resistant bacteria and antibiotic resistance determinants in the environment.

We used culture-independent approaches to determine the prevalence of ARGs and to examine how bacterial communities from biofilms and sediments respond to the discharge of WWTP effluents in the receiving river. ARGs and bacterial community composition in the upstream river were also analyzed to determine the contribution of wastewater discharge to antibiotic resistance in the downstream river samples. Biofilm and sediment were selected rather than water because they are substrates with a high bacterial density where the frequency of physical contacts between bacteria increase the possibility for horizontal gene transfer [15]. Actually, some studies have proposed aquatic biofilms as long-term reservoirs for ARGs in environment [9,16].

5.2 MATERIALS AND METHODS

5.2.1 STUDY SITE AND SAMPLING

The study was carried out in the Ter River upstream and downstream of the Ripoll WWTP discharge. This river is the water supply for most cities in Catalonia. The WWTP has a primary and secondary treatment operating with conventional activated sludge and was planned for a population of 45,000 inhabitants. It receives an average daily flow about 8000 m3 made up primarily of domestic wastewater and a small amount of industrial and hospital wastewater, which are not pre-treated. Samples were obtained in June 2010, the end of the spring season, when water flow is maximum. Biofilm and sediment samples were collected in duplicate at the WWTP discharge point and at 100 m upstream and downstream of the WWTP. Both sample types were collected manually, scraping the surface of submerged stones for epilithic biofilm and collecting the top layer (0-5 cm) of sediment. Water samples for antibiotics quantification were taken from influent and effluent of the WWTP. All samples were stored on ice and transported to the laboratory for immediate processing.

5.2.2 ETHICS STATEMENT

Permission for the WWTP samples was granted by the Ripoll Treatment Plant (Girona, Spain), specifically Angel Maderiano de Pastor (Supervisor, Wastewater Treatment Division). No specific permits or permissions were required for the samples collected in the Ter River.

5.2.3 ANTIBIOTICS QUANTIFICATION

To determine the efficiency of WWTP on antibiotics removal, some of these substances were quantified in WWTP influent and effluent. Quantification of antibiotics was carried out following the analytical methodology previously described [17]. Briefly, water samples were filtered through 1 μm glass microfibre filters followed by 0.45-μm nylon membrane filters (Whatman Maidstone, UK). Target compounds were extracted by solid-phase extraction using Oasis HLB cartridges (60mg, 3ml; Waters, Milford, MA, USA). Cartridges were loaded with 200 mL of water samples and a Baker vacuum system (J.T. Baker, Deventer, The Netherlands) was used to preconcentrate samples. The extracts were evaporated under a gentle nitrogen stream and reconstituted with 1 mL of methanol/water (25:75, v/v). Extracts were then analyzed by high-performance liquid chromatography tandem mass spectrometry using an Agilent HP 1100 HPLC (Agilent Technologies, Palo Alto, CA, USA) connected with a QTRAP hybrid triple quadrupole-linear ion trap mass spectrometer (Applied Biosystems/ MDS SCIEX, Foster City, CA, USA).

5.2.4 DNA EXTRACTION

Biofilm and sediment samples were homogenized in phosphate-buffered saline solution (PBS; 10 mM sodium phosphate, 150 mM sodium chloride, pH 7.2) and the supernatants were then resuspended in lysis buffer (20 mM Tris-Cl, pH 8.0; 2 mM sodium EDTA; 1.2 % Triton X-100; and 20 mg/ml lysozyme). Genomic DNA was extracted using a standard phenol-chloroform method and the final concentration and purity were determined using a NanoDrop spectrophotometer (Thermo Scientific, Wilmington, DE, USA). DNA was extracted in duplicate from each independent sample, obtaining 4 analytical replicates, and all DNA samples were stored at -20 °C until further analysis.

5.2.5 QUANTIFICATION OF ANTIBIOTIC RESISTANCE GENES

Quantitative PCR was used to quantify eleven genes encoding resistance to the main antibiotic families used for treating human and animal infections such as beta-lactams (bla_{TEM}, bla_{CTX-M} and bla_{SHV}), fluoroquinolones (qnrA, qnrB and qnrS), tetracyclines [tet(O) and tet(W)], sulfonamides [sul(I) and sul(II)] and macrolides [erm(B)]. The 16S ribosomal RNA (rRNA) gene was also analyzed to quantify the total bacterial load and to normalize the abundance of ARGs in the collected samples. All qPCR assays were performed on an Mx3005P system (Agilent Technologies) using SYBR Green detection chemistry. Each reaction was carried out in a total volume of 30 µl, containing 1 µl of template, the corresponding concentration of each primer (from 0.2 to 0.6 µM) and 2× Brilliant III Ultra Fast QCPR Master Mix (Stratagene, La Jolla, CA, USA), except for the bla_{TEM} gene, which was amplified using the SYBR® Green Master Mix (Applied Biosystems). Primers and thermal cycling conditions for each gene are given in supplementary material, Table S1. In all cases, DNA extracted from samples was diluted 10- and 100-fold and positive controls were spiked with our DNA samples in order to screen for PCR inhibition by environmental matrices. Moreover, after the PCR a dissociation curve was constructed in the range of 60°C to 95°C to verify the specificity of the amplified products.

Standard curves were generated using known quantities of cloned target genes. Briefly, amplicons from positive controls were ligated into pCR2.1-TOPO cloning vectors (Invitrogen, Carlsbad, CA, USA) and transformed into *Escherichia coli* competent cells following the manufacturer's protocol (Invitrogen). Plasmids were extracted using the PureLink Plasmid kit (Invitrogen), and the concentration was determined using a Nanodrop spectrophotometer (Thermo Scientific). Copy number was then calculated as described previously [18]. Tenfold serial dilutions of plasmid DNA were amplified over seven orders of magnitude and in triplicate to generate a standard curve for each qPCR assay.

Mean values (copy number of each ARG) of four analytical replicates for each sample were compared using the one-way analysis of variance (ANOVA) or Kruskal-Wallis test as appropriate, because most data were not normally distributed. Data were analyzed using SPSS for Windows version 17.0 (SPSS, Chicago, IL, USA).

5.2.6 SEQUENCE ANALYSIS AND PHYLOGENETIC CLASSIFICATION

DNA extraction replicates from each sample were pooled and submitted to the Research and Testing Laboratory (Lubbock, TX, USA) for tag-pyrosequencing. Samples were amplified with primers 27F (3'-GAG TTT GAT CNT GGC TCAG-5') and 519R (3'-GTN TTA CNG CGG CKG CTG-5'), and the amplicons were sequenced using Roche 454 GS-FLX Titanium technology [19]. Sequences were quality trimmed using the MOTHUR software package [20]. Briefly, we removed sequences that did not perfectly match the PCR primer at the beginning of a read, sequences that contained more than one ambiguous base, sequences having a homo-polymer stretch longer than 8 bp, and sequences with an average quality score below 30. We also included only the first 250 bp after the proximal PCR primer, because the quality of sequences degrades beyond this point. Then, sequences were aligned using the SILVA reference database [21] and potential chimeric sequences were detected and removed by using chimera.uchime incorporated into MOTHUR. Qualified sequences were assigned to operational taxonomic units (OTUs) based on a 97% sequence similarity. The Shannon diversity index (H') and the Chao1 richness estimator were also calculated. The Bray-Curtis distance, which incorporates both membership and abundance, was used to compare beta diversity among samples. The parsimony test, as implemented by MOTHUR, was used to assess whether two or more communities have the same structure. A Bonferroni correction was applied to adjust the significance level for multiple pairwise comparisons ($p \leq 0.05/15$ [0.0033]). The Ribosomal Database Project (RDP) pipeline and Classifier function were used to align and assign identities at a confidence threshold of 50% [22]. The sequences from this study have been deposited in the NCBI Short Read Archive under accession number SRA067245.

5.3 RESULTS

5.3.1 ANTIBIOTIC CONCENTRATIONS

Antibiotic compounds, such as clarithromycin, sulfamethoxazole, trime-throprim, metronidazole and ciprofloxacin were detected in WWTP influent and effluent samples at concentrations ranging from 20.8 to 913 ng/L (Table 1). Although concentrations found in the WWTP influent were

TABLE 1. Concentrations of antibiotics determined in WWTP influent and effluent water samples.

Antibiotic	Influent (ng/L)	Effluent (ng/L)
Clarithromycin	181.9	166.0
Trimethoprim	22.0	20.8
Metronidazole	161.0	43.3
Sulfamethoxazole	136.0	57.8
Ciprofloxacin	913.0	231.0

higher than those in the effluent, relatively high levels of antibiotic compounds were detected in treated water from the WWTP.

5.3.2 QUANTIFICATION OF ARGS

ARGs, including qnrA, qnrB, qnrS, bla_{TEM}, bla_{SHV}, bla_{CTX-M}, erm(B), sul(I), sul(II), tet(O) and tet(W), and the 16S rRNA gene were quantified by qPCR in the biofilm and sediment samples. High R2 values (average 0.995) and high efficiencies (from 95 to 103%) were obtained over at least 5 orders of magnitude in all qPCR assays, indicating the validity of these quantifications (data not shown). Results revealed that the total copy numbers of bacterial 16S rRNA genes were consistent in all samples and ranged from 1.45×10^9 to 1.21×10^{11} copy numbers per gram. Relative concentrations of ARGs (normalized to the 16S rRNA gene copy number) are shown in Figure 1. From this figure it can be seen that, except qnrA and qnrB, all ARGs were detected in the samples analyzed. It is noteworthy that relative abundances of the qnrB, qnrS, bla_{TEM}, bla_{SHV}, erm(B), sul(I), sul(II), tet(O) and tet(W) genes were significantly higher (p<0.05) in the downstream biofilm samples than those found in the upstream samples. Regarding

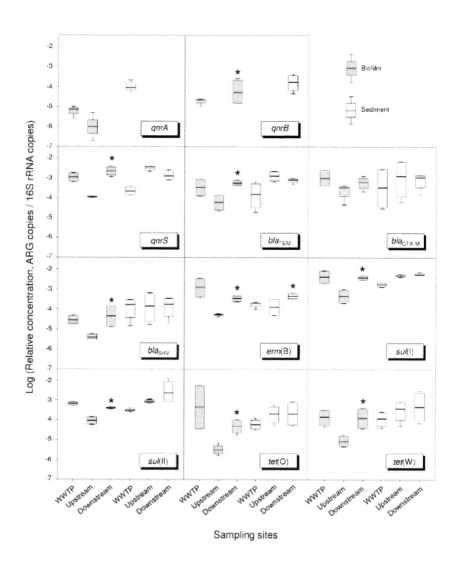

Figure 1. Relative concentration of ARGs in biofilm and sediment samples. Within the box plot chart, the crosspieces of each box plot represent (from top to bottom) maximum, upper-quartile, median (black bar), lower-quartile, and minimum values. An asterisk (*) denotes a statistically significant difference between upstream and downstream sites (P<0.05).

sediment samples, no significant differences in relative concentrations of ARGs were observed among them, except for the erm(B) gene.

5.3.3 BACTERIAL COMMUNITY COMPOSITION

A total of 77,056 reads from biofilm and sediment samples were obtained after quality trimming and filtering the initial reads. The library size of each sample was then normalized to the smallest number of sequences obtained from biofilm and sediment samples (4,328 and 7,587 sequences, respectively) in order to minimize any bias due to the difference in the total number of sequences. The number of OTUs observed at a 97% taxonomic cutoff ranged from 262 (in the upstream biofilm) to 2,527 (in sediment from the WWTP discharge point). Shannon diversity index and Chao richness estimators were also determined (Table 2), demonstrating that the sediment samples had a higher bacterial diversity and richness than the biofilm samples.

To compare the bacterial community structure and determine the effect of WWTP discharges into the environment, we used the phylogeny-based parsimony test which showed a significant difference (p<0.001) in community structure between all analyzed samples. When the pairwise distances for all samples were calculated using the Bray-Curtis distance metric, the results revealed the impact of the WWTP effluents on the bacterial community structure in the sediment of receiving river, as visualized by the terminal branch lengths (Figure 2).

Phylogenetic classification of sequences was determined using the RDP Classifier tool with a bootstrap cutoff of 50% (Figure 3). Overall taxonomic characterization of the bacterial community was conducted at the phylum level, and only Proteobacteria were classified at the class or order level. Biofilm sequences showed great differences between WWTP and river samples. Although the biofilm from the WWTP discharge point was dominated by *Cyanobacteria* and Proteobacteria, the most abundant groups in both upstream and downstream biofilms were Firmicutes. Gammaproteobacteria were mainly represented by the genera *Aeromonas* (16%) and *Acinetobacter* (8%) in the biofilm from the WWTP discharge point, whereas *Exiguobacterium* was the most predominant genus (data

TABLE 2. Measures of α diversity for the biofilm and sediment samples.

	Biofilm			Sediment		
	WWTP	Upstream	Downstream	WWTP	Upstream	Downstream
No. of sequences	4328	4328	4328	7587	7587	7587
OTUs	560	262	740	2527	1795	2202
H'	4.75	1.53	3.97	7.06	5.81	6.70
Chao1	1145	852	1988	6372	4378	5649

not shown), accounting for 85% of the observed OTUs in upstream samples and 46% in downstream samples. On the other hand, members of the Proteobacteria and Actinobacteria phyla were abundant in all sediment samples. Sequences from sediment samples were dominated by common genera such as *Aeromonas*, *Exiguobacterium*, *Piscinibacter*, *Pseudohodoferax*, *Acinetobacter* and *Pseudomonas*.

5.4 DISCUSSION

In this study we investigated the prevalence of eleven ARGs and the bacterial community composition in biofilm and sediment samples from a river influenced by a WWTP. It has been some years since Iwane and colleagues [23] showed that the increase of antibiotic resistant bacteria in the Tama River was associated with WWTP effluent discharges. Since then, a wide range of genetic methods have been developed and some culture-independent studies have been performed to explore the impact of antibiotics in the environment; however, most of these studies have been limited to a few ARGs [24–26].

Although WWTPs efficiently reduce high nutrient concentrations from raw sewage, our study demonstrates, to a certain extent, that antibiotics are

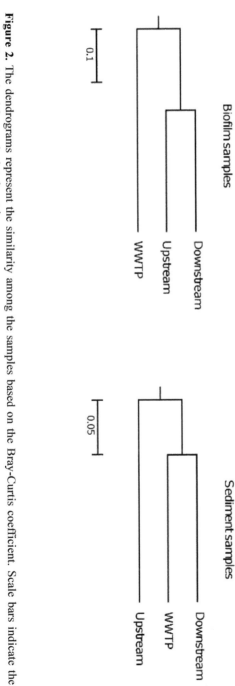

Figure 2. The dendrograms represent the similarity among the samples based on the Bray-Curtis coefficient. Scale bars indicate the similarity obtained from calculated matrices.

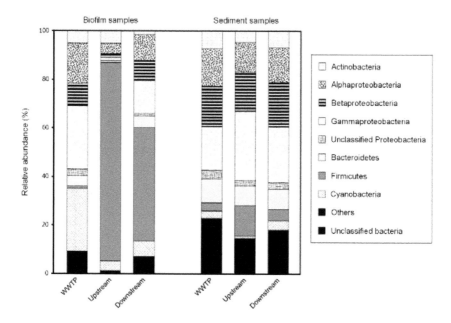

Figure 3. Relative abundance of major bacterial lineages (phyla; and classes for Proteobacteria) found in the biofilm and sediment samples.

not completely degraded during wastewater treatment. Previous studies have shown that antibiotic levels in treated wastewater are typically in the nanogram per liter range [15] and WWTP effluents are the major pathway for pharmaceuticals to reach the aquatic environment [27,28]. Effluents are diluted once they reach the river, and even though antibiotic concentrations in the environment are low, in general below the minimum inhibitory concentration for most sensitive bacteria, they may still exert a selection pressure [29,30] and impact on the microbial community [31].

Regarding ARG concentrations, we identified nine of the eleven ARGs studied in all samples analyzed. Detection of these genes in upstream biofilm and sediment samples supports the idea of an existing background level of antibiotic resistance naturally occurring in the environment [10]. However, other anthropogenic activities in the river upstream, such as livestock rearing, may have perturbed the bacterial community. We also report that there was a significant increase in the relative concentration for almost all ARGs studied in the biofilm samples after the WWTP effluent discharges. This is consistent with the observations of Engemann et al. [9] which suggested the migration of ARGs from the water column to biofilms. Our findings thus, together with those previously reported, suggest that biofilms could be considered as good indicators of antibiotic resistance acquisition. Moreover, our results reinforce the view that environmental compartments directly impacted by anthropogenic activities, such as wastewater discharges, show a higher concentration of ARGs [8,24,32].

Previous studies have also suggested that aquatic sediments could be important reservoirs for ARGs due to their ability to retain antibiotic compounds as well as representing an important environmental matrix within which horizontal gene transfer can occur [33]. LaPara et al. [32] detected ARGs encoding resistance to tetracyclines in water and sediments samples in locations influenced by WWTP effluents. Although ARGs were found in most of the sediment samples analyzed, we could not determine the effect of WWTP in the receiving river as some ARG levels in the discharge point had similar values in terms of relative concentration to those found in the upstream and downstream sediment samples.

We also investigated the potential effect of WWTP effluents on bacterial communities associated with the receiving river by using 454-pyrosequencing technology. Based on the Chao1 and Shannon indices, the bacterial communities in the sediment samples had higher richness and diversity than those in the biofilm samples. This could be because, although biofilms can be composed of multiple microbial species, they may also be dominated by a few genera in the primary colonization phase [34].

It has also been reported that sediments from aquatic environments have a complex and dynamic community of microorganisms [35]. In this study, the bacterial communities in the biofilm and sediment samples from the WWTP discharge point and downstream river had a higher diversity than those in the upstream samples, suggesting that WWTP effluent discharges may have promoted bacterial growth by supplying nutrients [36].

Despite the high degree of similarity among the bacterial communities from river and WWTP sediment samples at the phylum level, differences with respect to community structure and composition were significant at the genus level (97% similarity) as revealed by the phylogeny-based parsimony test. Similarly, Kristiansson et al. [6] found differences in the distribution of bacterial genera between the upstream and downstream sediments from an Indian river influenced by WWTP effluents. Moreover, our results also agreed with the observation of Kristiansson et al. [6] that all sediment communities were dominated by Proteobacteria, Bacteroidetes and Firmicutes. With regard to the biofilm samples, differences in community structures were evident even at the phylum level. *Cyanobacteria* and Proteobacteria were dominant in the WWTP discharge point, whereas Firmicutes was the dominant group in both upstream and downstream samples. A high representation of *Cyanobacteria* was found in biofilm samples from the WWTP, which may be a consequence of light intensity, as the WWTP effluent canal is less deep than the river, promoting the growth of phototrophic microorganisms. Concerning the relationship between changes in ARG concentrations and bacterial community composition, we found a higher proportion of Gammaproteobacteria in downstream biofilm samples compared with upstream samples, which could explain the increase of ARGs as several members of this class harbour ARGs [37,38]. A computer search using the Antibiotic Resistance Genes Database [39] also revealed that most of the bacterial species harboring ARGs belong to this class. In this sense, the similarity among the proportion of Gammaproteobacteria in bacterial communities from sediment samples could be related with the similar values of ARGs found in the different points analysed.

Additionally, we could detect and estimate the number of bacterial genera in each sample, which have been previously identified as harbouring ARGs. The genus *Exiguobacterium*, which was the dominant OTU in biofilm river samples and also appeared in sediment samples, has been recently characterized as a carrier of some ARGs encoding resistance to beta-lactams and sulphonamides [40]. The genus *Aeromonas*, which was detected in all samples, has been widely studied because most of ARGs we analyzed have been detected in several species of this genus [41,42]. Moreover, members of the genus Acinetobacter, which were present in high percentages in biofilm and sediment samples, have been also described as multidrug-resistant microorganisms encoding resistance to beta-lactams, aminoglycosides, fluoroquinolones and carbapenems [43]. Given this, microorganisms belonging to these genera may have contributed to the occurrence, spread and persistence of ARGs.

In conclusion, eleven different ARGs encoding resistance to the most important antibiotic families were analyzed using a culture-independent method, which contributes to a better understanding on the spread of antibiotic resistance in the environment. Of special concern is that our findings, together with reports from other settings, demonstrate that WWTP discharges may increase the prevalence of ARGs and bacterial community composition of the receiving river. However, further research is needed to evaluate if the increase of ARGs in aquatic ecosystems is due to the release of resistant bacteria from WWTP or due to antibiotics discharged in their effluents promoting horizontal gene transfer once they reach the river.

REFERENCES

1. McCoy E (1954) Changes in the host flora induced by chemotherapeutic agents. Annu Rev Microbiol 8: 257–272. doi:10.1146/annurev.mi.08.100154.001353. PubMed: 13198107.
2. Leeb M (2004) Antibiotics: a shot in the arm. Nature 431: 892–893. doi:10.1038/431892a. PubMed: 15496888.
3. Aminov RI, Mackie RI (2007) Evolution and ecology of antibiotic resistance genes. FEMS Microbiol Lett 271: 147–161. doi:10.1111/j.1574-6968.2007.00757.x. PubMed: 17490428.
4. Huerta B, Marti E, Gros M, López P, Pompêo M et al. (2013) Exploring the links between antibiotic occurrence, antibiotic resistance, and bacterial communities in water supply reservoirs. Sci Total Environ 456-457: 161–170. doi:10.1016/j.scitotenv.2013.03.071. PubMed: 23591067.
5. Di Cesare A, Luna GM, Vignaroli C, Pasquaroli S, Tota S et al. (2013) Aquaculture can promote the presence and spread of antibiotic-resistant enterococci in marine sediments. PLOS ONE 8: e62838. doi:10.1371/journal.pone.0062838. PubMed: 23638152.
6. Kristiansson E, Fick J, Janzon A, Grabic R, Rutgersson C et al. (2011) Pyrosequencing of antibiotic-contaminated river sediments reveals high levels of resistance and gene transfer elements. PLOS ONE 6: e17038. doi:10.1371/journal.pone.0017038. PubMed: 21359229.

7. Baquero F, Martínez JL, Cantón R (2008) Antibiotics and antibiotic resistance in water environments. Curr Opin Biotechnol 19: 260–265. doi:10.1016/j.copbio.2008.05.006. PubMed: 18534838.
8. Pruden A, Pei R, Storteboom H, Carlson KH (2006) Antibiotic resistance genes as emerging contaminants: studies in northern Colorado. Environ Sci Technol 40: 7445–7450. doi:10.1021/es0604131. PubMed: 17181002.
9. Engemann CA, Keen PL, Knapp CW, Hall KJ, Graham DW (2008) Fate of tetracycline resistance genes in aquatic systems: Migration from the water column to peripheral biofilms. Environ Sci Technol 42: 5131–5136. doi:10.1021/es800238e. PubMed: 18754359.
10. Allen HK, Donato J, Wang HH, Cloud-Hansen KA, Davies J et al. (2010) Call of the wild: antibiotic resistance genes in natural environments. Nat Rev Microbiol 8: 251–259. doi:10.1038/nrmicro2312. PubMed: 20190823.
11. Zhang K, Farahbakhsh K (2007) Removal of native coliphages and coliform bacteria from municipal wastewater by various wastewater treatment processes: implications to water reuse. Water Res 41: 2816–2824. doi:10.1016/j.watres.2007.03.010. PubMed: 17449083.
12. Munir M, Wong K, Xagoraraki I (2011) Release of antibiotic resistant bacteria and genes in the effluent and biosolids of five wastewater utilities in Michigan. Water Res 45: 681–693. doi:10.1016/j.watres.2010.08.033. PubMed: 20850863.
13. Mokracka J, Koczura R, Kaznowski A (2012) Multiresistant Enterobacteriaceae with class 1 and class 2 integrons in a municipal wastewater treatment plant. Water Res 46: 3353–3363. doi:10.1016/j.watres.2012.03.037. PubMed: 22507248.
14. LaPara T, Burch T (2012) Municipal wastewater as a reservoir of antibiotic resistance. In: PL KeenM. Montforts. Antimicrobial Resistance in the environment. Hoboken, NJ: Wiley-Blackwell. pp. 241–250.
15. Janzon A, Kristiansson E, Larsson DGJ (2012) Environmental microbial communities living under very high antibiotic selection pressure. In: PL KeenM. Montforts. Antimicrobial Resistance in the environment. Hoboken, NJ: Wiley-Blackwell. pp. 483–501.
16. Zhang XX, Zhang T, Fang HH (2009) Antibiotic resistance genes in water environment. Appl Microbiol Biotechnol 82: 397–414. doi:10.1007/s00253-008-1829-z. PubMed: 19130050.
17. Gros M, Petrović M, Barceló D (2009) Tracing pharmaceutical residues of different therapeutic classes in environmental waters by using liquid chromatography/ quadrupole-linear ion trap mass spectrometry and automated library searching. Anal Chem 81: 898–912. doi:10.1021/ac801358e. PubMed: 19113952.
18. Marti E, Balcázar JL (2013) Real-time PCR assays for quantification of qnr genes in environmental water samples and chicken feces. Appl Environ Microbiol 79: 1743–1745. doi:10.1128/AEM.03409-12. PubMed: 23275512.
19. Acosta-Martinez V, Dowd S, Sun Y, Allen V (2008) Tag-encoded pyrosequencing analysis of bacterial diversity in a single soil type as affected by management and land use. Soil Biol Biochem 40: 2762–2770. doi:10.1016/j.soilbio.2008.07.022.
20. Schloss PD, Westcott SL, Ryabin T, Hall JR, Hartmann M et al. (2009) Introducing mothur: open-source, platform-independent, community-supported software

for describing and comparing microbial communities. Appl Environ Microbiol 75: 7537–7541. doi:10.1128/AEM.01541-09. PubMed: 19801464.

21. Pruesse E, Quast C, Knittel K, Fuchs BM, Ludwig W et al. (2007) SILVA: a comprehensive online resource for quality checked and aligned ribosomal RNA sequence data compatible with ARB. Nucleic Acids Res 35: 7188–7196. doi:10.1093/nar/gkm864. PubMed: 17947321.

22. Wang Q, Garrity GM, Tiedje JM, Cole JR (2007) Naive Bayesian classifier for rapid assignment of rRNA sequences into the new bacterial taxonomy. Appl Environ Microbiol 73: 5261–5267. doi:10.1128/AEM.00062-07. PubMed: 17586664.

23. Iwane T, Urase T, Yamamoto K (2001) Possible impact of treated wastewater discharge on incidence of antibiotic resistant bacteria in river water. Water Sci Technol 43: 91–99. PubMed: 11380211.

24. Pei R, Kim SC, Carlson KH, Pruden A (2006) Effect of river landscape on the sediment concentrations of antibiotics and corresponding antibiotic resistance genes (ARG). Water Res 40: 2427–2435. doi:10.1016/j.watres.2006.04.017. PubMed: 16753197.

25. Chen J, Yu Z, Michel FC, Wittum T, Morrison M (2007) Development and application of real-time PCR assays for quantification of erm genes conferring resistance to macrolide-lincosamides-streptogramin B in livestock manure and manure management systems. Appl Environ Microbiol 14: 4407–4416. doi: 10.1128/aem.02799-06

26. Li J, Wang T, Shao B, Shen J, Wang S et al. (2012) Plasmid-mediated quinolone resistance genes and antibiotic residues in wastewater and soil adjacent to Swine feedlots: Potential transfer to agricultural lands. Environ Health Perspect 120: 1144–1149. doi:10.1289/ehp.1104776. PubMed: 22569244.

27. Petrovic M, Gonzalez S, Barceló D (2003) Analysis and removal of emerging contaminants in wastewater and drinking water. Trends Anal Chem 22: 685–696. doi:10.1016/S0165-9936(03)01105-1.

28. Gros M, Rodríguez-Mozaz S, Barceló D (2012) Fast and comprehensive multi-residue analysis of a broad range of human and veterinary pharmaceuticals and some of their metabolites in surface and treated waters by ultra-high-performance liquid chromatography coupled to quadrupole-linear ion trap tandem mass spectrometry. J Chromatogr A 1248: 104–121. doi:10.1016/j.chroma.2012.05.084. PubMed: 22704668.

29. Martínez JL (2008) Antibiotics and antibiotic resistance genes in natural environments. Science 321: 365–367. doi:10.1126/science.1159483. PubMed: 18635792.

30. Kümmerer K (2009) Antibiotics in the aquatic environment - A review - Part I. Chemosphere 75: 417–434. doi:10.1016/j.chemosphere.2008.11.086. PubMed: 19185900.

31. Davies J, Spiegelman GB, Yim G (2006) The world of subinhibitory antibiotic concentrations. Curr Opin Microbiol 9: 445–453. doi:10.1016/j.mib.2006.08.006. PubMed: 16942902.

32. LaPara TM, Burch TR, McNamara PJ, Tan DT, Yan M et al. (2011) Tertiary-treated municipal wastewater is a significant point source of antibiotic resistance genes into Duluth-Superior Harbor. Environ Sci Technol 45: 9543–9549. doi:10.1021/es202775r. PubMed: 21981654.

33. Taylor NG, Verner-Jeffreys DW, Baker-Austin C (2011) Aquatic systems: maintaining, mixing and mobilising antimicrobial resistance? Trends Ecol Evol 26: 278–284. doi:10.1016/j.tree.2011.03.004. PubMed: 21458879.

34. Sigee DC (2005) Freshwater Microbiology: Biodiversity and Dynamic Interactions of Microorganisms in the Aquatic Environment. Chichester, England: John Wiley & Sons Ltd.. p. 524.

35. Lu SY, Zhang YL, Geng SN, Li TY, Ye ZM et al. (2010) High diversity of extended-spectrum beta-lactamase-producing bacteria in an urban river sediment habitat. Appl Environ Microbiol 76: 5972–5976. doi:10.1128/AEM.00711-10. PubMed: 20639374.

36. Wakelin SA, Colloff MJ, Kookana RS (2008) Effect of wastewater treatment plant effluent on microbial function and community structure in the sediment of a freshwater stream with variable seasonal flow. Appl Environ Microbiol 74: 2659–2668. doi:10.1128/AEM.02348-07. PubMed: 18344343.

37. Li D, Yu T, Zhang Y, Yang M, Li Z, Liu M, Qi R (2010) Antibiotic resistance characteristics of environmental bacteria from an oxytetracycline production wastewater treatment plant and the receiving river. Appl Environ Microbiol 76: 3444–3451. doi:10.1128/AEM.02964-09. PubMed: 20400569.

38. Tamminen M, Virta M, Fani R, Fondi M (2012) Large-scale analysis of plasmid relationships through gene-sharing networks. Mol Biol Evol 29: 1225–1240. doi:10.1093/molbev/msr292. PubMed: 22130968.

39. Liu B, Pop M (2009) ARDB--Antibiotic Resistance Genes Database. Nucleic Acids Res 37: D443–D447. doi:10.1093/nar/gkn656. PubMed: 18832362.

40. Carneiro AR, Ramos RT, Dall'Agnol H, Pinto AC, de Castro Soares S et al. (2012) Genome sequence of Exiguobacterium antarcticum B7, isolated from a biofilm in Ginger Lake, King George Island, Antarctica. J Bacteriol 194: 6689–6690. doi:10.1128/JB.01791-12. PubMed: 23144424.

41. Cattoir V, Poirel L, Aubert C, Soussy CJ, Nordmann P (2008) Unexpected occurrence of plasmid-mediated quinolone resistance determinants in environmental Aeromonas spp. Emerg Infect Dis 14: 231–237. doi:10.3201/eid1402.070677. PubMed: 18258115.

42. Marti E, Balcázar JL (2012) Multidrug resistance-encoding plasmid from Aeromonas sp. strain P2G1. Clin Microbiol Infect 18: E366–E368. doi:10.1111/j.1469-0691.2012.03935.x. PubMed: 22725683.

43. Mak JK, Kim MJ, Pham J, Tapsall J, White PA (2009) Antibiotic resistance determinants in nosocomial strains of multidrug-resistant Acinetobacter baumannii. J Antimicrob Chemother 63: 47–54. PubMed: 18988680.

There is a supplemental file that is not available in this version of the article. To view this additional information, please use the citation on the first page of this chapter.

CHAPTER 6

Impact of Treated Wastewater Irrigation on Antibiotic Resistance in the Soil Microbiome

JOAO GATICA AND EDDIE CYTRYN

6.1 INTRODUCTION

The world is confronted by an ever increasing shortage of water, especially in arid and semiarid regions such as Africa, South Asia, Southern Europe, and the Middle East. In many of these regions, freshwater is not available for irrigation, and therefore, the reuse of treated or untreated wastewater is the sole water source for agriculture. However, standards are required to ensure safe use of wastewater and to avoid biological risks to the human population. In this context, the dissemination of antibiotic-resistant bacteria (ARB) and antibiotic-resistant genes (ARGs) from wastewater irrigation to natural soil and water environments is of public concern because it may contribute to global antibiotic resistance (AR). Here, we discuss the implications of wastewater reuse in agriculture and the impact of treated wastewater (TWW) irrigation on AR in the soil microbiome.

© *The Author(s) 2013.* Environmental Science and Pollution Research *June 2013, Volume 20, Issue 6, pp 3529-3538; 10.1007/s11356-013-1505-4. This article is distributed under the terms of the Creative Commons Attribution License.*

6.2 WASTEWATER REUSE IN AGRICULTURE: BENEFITS VS. RISKS

Mankind is currently confronted with one of the greatest challenges in its history: how to adequately use its limited freshwater resources. In this context, the challenge implicates the use of water for drinking, agriculture, and the preservation of fragile freshwater ecosystems (Postel 2000).

The reuse of TWW, especially in agriculture, is an appealing and practical solution for water scarcity that significantly relieves pressure on water resources (Toze 2006; Zhang and Liu 1989). Additionally, water reuse can alleviate the discharge of effluents into the environment, avoiding in this way the deterioration of freshwater ecosystems associated with eutrophication and algal blooms (Toze 2006). The use of TWW in agricultural irrigation has also been found to have additional agronomic benefits associated with soil structure and fertility. According to Kiziloglu et al. (2007), wastewater has a high nutritive value that may improve plant growth, reduce fertilizer application rates, and increase productivity of poor fertility soils. Diverse studies have indeed shown that TWW irrigation increases soil organic matter (Mañas et al. 2009; Jueschke et al. 2008; Kiziloglu et al. 2007) as well as the concentrations of different nutrients involved in plant growth such as nitrogen (N), phosphorus, iron, manganese, potassium, calcium, magnesium, and others (Akponikpe et al. 2011; Rezapour and Samadi 2011; Sacks and Bernstein 2011; Mañas et al. 2009; Gwenzi and Munondo 2008; Kim et al. 2007; Kiziloglu et al. 2007; Angin et al. 2005). Conversely, the use of TWW for irrigation can have detrimental effects on soil quality. These include increased salinity and decreased soil pH (Kiziloglu et al. 2007; Angin et al. 2005; Mohammad and Mazahreh 2003), as well as increased soil hydrophobicity (Tarchitzky et al. 2007; Graber et al. 2006).

Despite the obvious benefits of TWW irrigation, the human and environmental health implications of this process have opened a new controversial front in the public debate (Phung et al. 2011). Perhaps the most evident public health qualms are linked to the presence pathogens. Several studies have reported high count of total coliforms and fecal coliforms in crops irrigated with TWW (Akponikpe et al. 2011; Sacks and Bernstein 2011; Mutengu et al. 2007; Rai and Tripathi 2007), while others have detected bacterial pathogens such as *Salmonella, Streptococci, Clostridium, Shigella*, and *Vibrio* spp. (Mañas et al. 2009; Samie et al. 2009). Other human pathogens associated with wastewater reuse include helminthes, viruses, and protozoa (Carey et al. 2004; Caccio et al. 2003; Tree et al. 2003). The level of microbial contamination observed in recycled wastewater, soils, and crops depends on technical regulations based on national standards. In this context, posttreatment disinfection processes such as chlorination, ozonation, and UV radiation, shown to be successful treatments against microbial agents and pharmaceutical ingredients, can significantly reduce the risks associated with TWW irrigation (Hey et al. 2012; Martinez et al. 2011; Nikaido et al. 2010; Bernstein et al. 2008; An et al. 2007).

Irrigation of agricultural crops with recycled wastewater is also associated with several non-biological risk factors; perhaps the most significant being heavy metal contamination. In this context, different studies have shown the accumulation of heavy metals such as cadmium (Cd), nickel, chromium (Cr), lead, and other elements in soil and plants under wastewater irrigation regimen (Gupta et al. 2010; Bahmanyar 2008; Khan et al. 2008; Song et al. 2006; Wang et al. 2006; Mapanda et al. 2005). According to Gupta et al. (2010), among the heavy metals aforementioned, Cd and Cr are of greatest concern due to their high uptake rates in plants and their accumulation in tissue vegetal body parts; implicating a health hazard associated with the consumption of these heavy metal-contaminated vegetables over a long period of time.

A wide spectrum of persistent organic contaminants has been detected in soils irrigated with TWW, including polycyclic aromatic hydrocarbons, polychlorinated biphenyls, and organochlorine pesticides (Chen et al. 2011; Sun et al. 2009; Chen et al. 2005; Pedersen et al. 2003). In addition, there is a rising concern regarding the presence of emerging contaminants that include endocrine-disrupting chemicals and pharmaceutical and personal care products (PPCPs), a diverse collection of thousands of chemical substances that include prescription and over-the-counter therapeutic drugs, veterinary drugs, fragrances, and cosmetics (EPA 2012). Several studies have suggested that the behavior and possible accumulation of PPCPs in natural environments could have a potential impact on both soil and human health (Walker et al. 2012; Chen et al. 2011; Xu et al. 2009a). However, more significant than the direct toxicity of PPCPs themselves are the potential biological effects that these compounds may have on downstream ecosystems. Perhaps, the most crucial of these effects is augmentation of AR in environmental microbiomes due to selective pressure.

With the aim of reducing risks associated with wastewater reuse, the first national standard for wastewater reuse was created in Israel in 1953 (Tal 2006). Since then, many guidelines and quality standard have been designed and applied in both Israel and worldwide (Inbar 2007; Blumenthal et al. 2000), but none have proved to be universally applicable (Phung et al. 2011) and none have considered the potential environmental and public health impact of effluent-associated emerging contaminants and their biological implications.

6.3 PERSISTENCE OF PHARMACEUTICALS
IN EFFLUENTS AND TWW-IRRIGATED SOILS

Certain pharmaceuticals are only partially metabolized in the body and, therefore, significant levels of these compounds and their metabolites are excreted and transported to wastewater treatment plants (WWTPs) (Xu et al. 2007). The removal or inactivation of these compounds in the WWTP depends on both the specific technology applied in the sewage plant and on the physical and chemical properties of each pharmaceutical compound (Monteiro and Boxall 2010). Nonetheless, many pharmaceutical compounds and their metabolites persist in WWTPs and are released in effluents even when rigorous tertiary disinfection methods are applied (Jelic et al. 2011).

PCPPs observed in the environment include analgesics and anti-inflammatories, antibiotics, antidepressants, antiepileptics, antineoplastic agents, β-blockers, hormones, different class of metabolites, and other pharmaceuticals (Fatta-Kassinos et al. 2011b; Monteiro and Boxall 2010; Nikolaou et al. 2007). Specific PCPPs commonly detected in WWTPs effluent include ibuprofen, diclofenac, naproxen and ketoprofen (analgesic and anti-inflammatories), atenolol, metoprolol, and propanolol (β-blockers), clofibric acid, bezafibrate and gemfibrozil (lipid regulator agent), carbamazepine (anti-seizure, antiepileptic), caffeine (stimulant), and triclosan (bactericide), which additionally are found in crops irrigated with TWW (Verlicchi et al. 2012; Bondarenko et al. 2012; Fenet et al. 2012; Chen et al. 2011; Fatta-Kassinos et al. 2011; Sim et al. 2011; Xu et al. 2009b; Chefetz et al. 2008; Topp et al. 2008; Kinney et al. 2006; Pedersen et al. 2003, 2005). Many antibiotic compounds are not degraded or only partially degraded in the body and often unused drugs are directly disposed of down drains (Kummerer and Henninger 2003). Additionally, several antibiotic compounds are not fully degraded in the WWTP process and are therefore released via effluents into the environment where they can select for antibiotic-resistant bacteria (Kummerer and Henninger 2003). Thus, antibiotics are considered pseudo-persistent contaminants because they are continuously introduced into the environment (Shi et al. 2012). Antibiotic compounds consistently detected in WWTP effluents include azithromycin, clarithromycin, ciprofloxacin, erythromycin, norfloxacin, ofloxacin, sulfamethoxazole, and trimethoprim (Fatta-Kassinos et al. 2011b; Li and Zhang 2011; Sim et al. 2011; Ghosh et al. 2009; Watkinson et al. 2007). Additionally, some studies have detected antibiotics, such as erythromycin and tetracycline, in wastewater-irrigated soils (Chen et al. 2011; Kinney et al. 2006) and sulfamethoxazole and erythromycin have even been discovered in groundwater under land irrigated with wastewater effluents (Avisar et al. 2009; Siemens et al. 2008).

6.4 HUMAN HEALTH IMPLICATIONS OF AR AND THE EMERGENCE OF ENVIRONMENTAL AR RESERVOIRS

According to Baquero et al. (2009), "antibiotics are among the most successful medical inventions, alleviating human morbidity and mortality; however, since they were introduced for therapy nearly 60 years ago, bacteria have developed different strategies to avoid their activity". Antibiotics are regularly used for treating infections and protecting human and animal health. In addition, they are frequently used to promote animal growth and improve feed efficiency in aquaculture and farming (Davies and Davies 2010; Baquero et al. 2009; Binh et al. 2008; Dzidic et al. 2008). However, the rapid emergence of AR since the development of antibiotics in the 1940s is extremely alarming, and some have gone as far as to state that we are approaching a "post-antibiotic era" (Alanis 2005). To exercise its action, an antibiotic compound needs to enter the bacterial cell at inhibitory or lethal concentrations, remain stable, and finally locate and interact with its target (Jayaraman 2009). However, bacteria can avoid one or more of the previously described steps by antibiotic inactivation, target modification, efflux pumps, target bypass, and non-inheritable mechanisms such as persistence, biofilm production, and swarming (Jayaraman 2009; Dzidic et al. 2008).

Traditional study of bacterial AR has focused on isolation of clinically important resistant pathogens that proliferate in response to antibiotic treatment in nosocomial and community settings. However, a myriad of recent data suggests that environmental reservoirs are also strongly associated with the global proliferation of AR (Forsberg et al. 2012; D'Costa et al. 2006, 2011). Several studies have indicated that anthropogenic activities such as aquaculture, application of manure and biosolids to soil, and environmental discharge of WWTP effluents can contribute to expansion of these environmental AR reservoirs (Munir and Xagoraraki 2011; Davies and Davies 2010; Knapp et al. 2010; Binh et al. 2008), and these reservoirs can serve as "hotspots" for the spread of ARGs and ARB through food and water, with unsuspected consequences for human health (Martinez 2009b; Zhang et al. 2009).

ARGs are often carried on broad host range mobile genetic elements (MGEs) (Byrne-Bailey et al. 2011; Akiyama et al. 2010; Binh et al. 2008), which are of particular concern because these vectors may be disseminated from environmental hotspots through water and food webs into clinically relevant pathogenic or opportunistic bacteria (Van Meervenne et al. 2012; Jayaraman 2009). Benveniste and Davies (1973) first suggested a link between environmental and clinical AR after detecting high similarities between gentamicin-resistant enzymes from soil-associated Actinomycetes and enzymes that confer the same resistance in human pathogens such *Escherichia coli, Pseudomonas aeruginosa, Klebsiella pneumonia,* and *Staphylococcus aureus*. Although a direct transfer of genetic information from one to other was not demonstrated, the authors suggested an evolutionary relationship between these enzymes in non-phylogenetically related bacteria. Highly conclusive evidence demonstrating the horizontal transfer of ARGs between soil and clinical isolates was recently presented by Forsberg et al. (2012), who applied functional metagenomics and high-throughput screening to large collections of soil isolates. The authors detected a wide array of functionally diverse ARGs in the soil isolates with perfect nucleotide identity to human pathogens. The data suggests that horizontal gene transfer (HGT) of AR genes between soil and clinical environments not only occurs, but appears to be a relatively common phenomenon.

6.5 RELEASE OF ARB AND ARGS FROM WWTPS

Antibiotic compounds have been found to select for resistance even when concentrations are orders of magnitude below clinical breakpoints (Gullberg et al. 2011), and therefore, persistent antibiotics such as erythromycin, tetracycline, or sulfamethoxazole may select for antibiotic resistance in downstream terrestrial and aquatic environments. In addition, heavy metals and biocides such zinc (Peltier et al. 2010) and triclosan (Ciusa et al. 2012; Aiello et al. 2005) can promote AR in bacteria in a process known as cross-selection (Martinez 2009a). In this context, WWTPs are a favorable environment for propagation of AR because they assemble high concentrations of bacteria (often in biofilms), nutrients, and antimicrobial agents (LaPara et al. 2011; Novo and Manaia 2010; Xi et al. 2009; DaCosta et al. 2006).

In the last years, more attention has been put on the role of WWTPs as reservoirs of ARB and ARGs, and different techniques have been used to assess their presence in different stages of wastewater treatment, as well as in downstream environments. Gao et al. (2012) found high concentrations of tetracycline (tetO and tetW) and sulfonamide (sulI) resistance genes (9.12×10^5–1.05×10^6 gene copy #/mL) and high levels of ARB (1.05×10^1–3.09×10^3 CFU/mL) in the final effluent of a wastewater treatment plant in Michigan. An additional study found significant levels of ARGs (tetO, tetW, and sulI) and ARB in raw sewage, effluent, and biosolid samples of a WWTP, although increased levels of wastewater treatment significantly reduced the amount of ARGs and ARBs in the treated effluent, especially when more advanced technologies such as membrane biological reactors were used (Munir et al. 2011). Previous reports have found other important ARGs such as ampC and vanA that confer resistance to ampicillin and vancomycin, respectively, in wastewater samples collected from five municipal treatment plants; additional studies found the methicillin-resistant gene mecA, in samples taken from clinical wastewater (Volkmann et al. 2004). Other studies have assessed AR in mobile elements such as plasmids, transposons, and integrons in WWTPs. In this context, a plasmid metagenome analysis from the final effluent of a WWTP in Germany revealed 140 clinically relevant ARGs including genes conferring resistance to aminoglycosides, β-lactams, chloramphenicol, fluoroquinolones, macrolides, rifampicin, tetracycline, trimethoprim, and sulfonamide as well as multidrug efflux and multidrug resistance genes (Szczepanowski et al. 2009). Screening of multiresistant Enterobacteriaceae isolates from different stages of a municipal WWTP in Poland revealed the presence of integron-positive isolates in the final effluent; all of whom were multiresistant to at least three different antimicrobials (Mokracka et al. 2012). A number of recent studies in Minnesota and Colorado detected high levels of genes encoding resistance to tetracycline (tetO, tetX, and tetW in Minnesota and tetO and tetW in Colorado) and the class 1 integron genes (intI1) in both tertiary treated effluents and in rivers receiving these effluents, demonstrating the transfer capacity of ARB and ARGs from WWTPs to aquatic environments (LaPara et al. 2011; Pruden et al. 2006). This phenomenon was established in a recent comprehensive study by Czekalski et al. (2012) who found that ARGs and ARB released from WWTP effluents were disseminated into Lake Geneva resulting in significant proliferation of AR levels in both the water column and in sediments that were proximal to the point of effluent infiltration.

6.6 NATIVE AR IN SOIL BACTERIA

Although human-associated activities may influence levels of AR in environmental microbiomes, it is becoming increasingly clear that ARB and ARGs are often highly prevalent in natural or undisturbed soils, assumedly because soil microorganisms are the main producers of clinical antibiotic compounds (Aminov and Mackie 2007). In this context, D'Costa et al. (2006), indicated that the soil could serve as an under-recognized reservoir for antibiotic resistance that has already emerged or has the potential to emerge in clinically important bacteria. The authors coined the term "resistome" to describe the collection of all the ARGs and their precursors in a defined natural environment. Several recent studies have strengthened the resistome hypothesis through the detection of ARGs in a wide array of undisturbed environments. For instance, soil bacterial genes encoding β-lactamases were detected in an undisturbed soil in Alaska (Allen et al. 2009), and recently, multiresistant bacteria were isolated from a deep and undisturbed cave in the Carlsbad Caverns National Park (USA) (Bhullar et al. 2012). Interesting, three *Streptomyces* isolates from this cave were highly resistant to daptomycin (MIC ≥ 256 µg/mL) which is employed as a last resort antibiotic in the treatment of drug-resistant Gram-positive pathogens (Bhullar et al. 2012). Perhaps, the most significant indication that AR is a natural phenomenon was provided by recent metagenomic-based analysis of authenticated ancient DNA from 30,000-year-old permafrost sediments in northern Canada. The authors detected a highly diverse collection of genes encoding resistance to β-lactam, tetracycline, and glycopeptide antibiotics and most surprisingly identified a vancomycin resistance element (VanA) that was highly similar in structure and function to modern variants of the gene vanA identified in clinical pathogens (D'Costa et al. 2011). The results detailed above strongly suggest that native AR in the soil microbiome has been underestimated until now and more studies are required to understand the real scope of the this phenomenon in the soil microbiome.

6.7 IMPACT OF WASTEWATER IRRIGATION ON THE SOIL MICROBIOME

As previously indicated, TWW irrigation is crucial for agriculture, especially in arid and semiarid areas of the world. However, based on the above-cited studies, TWW may harbor antibiotic compounds and metabolites as well as ARGs and ARB, which could enhance AR in the soil microbiome.

6.7.1 IMPACT OF TWW IRRIGATION ON SOIL MICROBIOME DIVERSITY AND ACTIVITY

Microbial activity, a key indicator of soil quality, is often used to assess the impact of anthropogenic and agronomic practices on soil vitality. Several studies have addressed the effect of TWW irrigation on soil microbial activity. Filip et al. (1999, 2000) reported higher microbial activity, measured by the activity of β-glucosidase, β-acetylglucosaminidase, proteinase, and phosphatase, in two soils irrigated with primary effluent for 100 years, relative to soils that were never irrigated. Interestingly, when these soils were left un-irrigated for 20 years, their microbial activities returned to levels characteristic of the nonirrigated soils. Additionally, the application of TWW significantly stimulated the development of copiotrophic bacteria and fungi, whereas the original soil microbiome was dominated by oligotrophic bacteria (require less nutrient to growth). The authors also detected strong increases in microbial biomass in soils irrigated with TWW. A study by Hidri et al. (2010) also found that long-term irrigation with TWW resulted in increased soil microbial abundance and TWW irrigation induced a particular composition of the bacterial and fungal communities. However, the magnitude and specificity of these changes were significantly correlated to the duration of irrigation. Additional studies conducted under different conditions with different soil types and TWW irrigation regimens also showed higher microbial activity in soils irrigated with TWW, but when the irrigation was suspended, the microbial activity returned to nonirrigated or freshwater-irrigated soil levels (Adrover et al. 2012; Elifanz et al. 2011; Meli et al. 2002), demonstrating the high resilience of the soil microbiome.

Microbial diversity and community structure are considered to be excellent indicators of soil health, and these methods have been implemented to assess the impact of anthropogenic activities on soil microbiota. In this context, Oved et al. (2001) and Ndour et al. (2008) evaluated the fate of ammonia-oxidizing bacteria (AOB) in two soils under different wastewater irrigation regimens (short- and long-term irrigated soils, respectively). In both cases, the results suggested that wastewater irrigation produces shifts in AOB population in the soil, as compared to soil irrigated with freshwater or groundwater, respectively. By applying denaturing gradient gel electrophoresis (DGGE) and subsequent cloning of excised DGGE bands, the Oved study revealed that the AOB population in wastewater-irrigated soils is dominated by *Nitrosomonas*-like strains, while in freshwater-irrigated soils, the AOB populations were dominated by *Nitrosospira*-like strains. However, despite shifts in the microbial community between wastewater- and freshwater-irrigated soils, no apparent shifts were observed in community function. Ndour et al. (2008) also saw no differences between the two treatments in microbial biomass or microbial activity, measured as FDA activity; presumably, due to the similar levels of organic C and N among the treatments. In contrast to the above-cited studies, Truu et al. (2009) reported that short-term municipal wastewater irrigation of a short-rotation willow coppice weakly affected soil chemical, microbiological, and biochemical properties. However, these changes were attributed to the willow growth rather than wastewater irrigation.

In general, the studies detailed above indicate that TWW-irrigated soils are characterized by a certain tendency for higher microbial activity, higher microbial biomass, and higher resilience than concurrent freshwater-irrigated soils. However, it should be noted that the final effect of TWW irrigation on the soil microbiome will depend on the compounds present in the TWW and their concentration, the duration of irrigation, and the properties of the soil irrigated. Based on currently available data, it appears that TWW-induced changes in microbial activity and community composition are primarily associated with increased salinity and levels of dissolved organic matter relative to and freshwater-irrigated soils. Although residual levels of PPCPs and antibiotics have been detected in TWW-irrigated soils, currently available methodologies are unable to link the presence of these compounds to soil microbial activity or community composition. To seek changes in the whole microbial activity as a result of soil disturbances, it may be advisable to observe the community dynamics of less abundant components of the soil microbiota because these taxonomic groups can be more sensitive than widespread groups in evidencing changes in the microbiome in response to soil disturbances (Gelsomino et al. 2006). The high abundance of AR in the Actinobacteria class may make this group a key target for assessing the impact of TWW irrigation on soil community composition.

6.7.2 IMPACT OF TWW IRRIGATION ON ARB AND ARG ABUNDANCE

One of the most significant contributions to the understanding of the effects of TWW irrigation on the magnitude of soil ARB and ARGs was recently published by Negreanu et al. (2012). In this study, four soils with different physicochemical properties were irrigated in tandem with freshwater or TWW and were assessed by standard culture-based isolation methods and culture-independent molecular analyses. Resistance to four clinically relevant antibiotic families—tetracycline, erythromycin, sulfonamide, and ciprofloxacin—was evaluated. The authors monitored six different ARGs that confer resistance to the abovementioned antibiotics, which had previously been detected in wastewater effluents. The genes chosen were qnrA gene that confers resistance to fluoroquinolones in Gram-negative bacteria, tetO linked to tetracycline resistance, sul(1) and sul(2) genes that confer resistance to sulfonamides, and erm(B) and erm(F) associated with resistance to macrolide, lincosamide, and streptogramin antibiotics. Surprisingly, our findings showed that the relative abundance of resistant isolates, and the levels of ARGs, was either identical or often even higher in freshwater-irrigated soils relative to the TWW-irrigated soils, despite significant loads of ARB and ARGs in the TWW that was used for irrigation. These results indicate that residual antibiotic concentrations associated with WWTP effluents do not seem to exert selective pressure that is significant enough to induce propagation of ARGs in TWW-irrigated soils and that TWW-associated ARB and ARGs do not persist in the irrigated soils. The authors concluded that the high numbers of resistant bacteria that enter the soils through TWW irrigation are not able to compete or survive in the soil environment and that the high levels of ARB and ARGs observed in both freshwater- and TWW-irrigated soils are predominantly associated with the native soil resistome. Preliminary bench-scale mesocosm experiments performed in our lab (Gatica unpublished), in which soils were subjected to an irrigation regime of dilute organic media amended with clinical concentrations of antibiotics in 250 ml plastic containers, showed no changes in microbial activity and very little changes in microbial community composition in relation to non-antibiotic-amended soils, supporting the notion that environmentally relevant antibiotic concentrations do not appear to significantly exert selective pressure on the soil microbiome.

Similar results were obtained in a recent study conducted in Arizona by McLain and Williams (2010), who compared AR profiles of Enterococcus isolates (screened against 16 antibiotic compounds) from soils irrigated for more than 20 years with either recharged municipal TWW or groundwater. The authors found high levels of AR in isolates in both TWW- and freshwater-irrigated soils, although AR patterns in the two soils differed. For example, isolates from the TWW-irrigated soils showed high resistance to daptomycin, and lincomycin, while isolates from the groundwater-filled pond were highly resistant to erythromycin, tetracycline, ciprofloxacin, and tylosin tartrate. Surprisingly, isolates from the freshwater soils showed higher levels of multiresistance than those isolated in the TWW-irrigated soils. Although the possibility of selective pressure due to low levels of antibiotic exposure from unknown sources cannot be rejected, the results reflect natural occurrence of AR in soils prior to TWW irrigation and suggest that ARB that entered the soil from the TWW did not survive in the soil environment. Supporting this idea, a recent study that assessed the survival of greywater- and TWW-associated bacterial pathogens in irrigated soils showed that while a high number of pathogens were present in TWW and greywater, no significant differences were observed between TWW/greywater-irrigated soils and freshwater-irrigated soil (Orlofsky et al. 2011).

6.8 OPEN QUESTIONS

When assessing the impact of TWW irrigation on AR in the soil microbiome, we should not ignore the mechanisms that govern the gene exchange at inter- and intra-species levels. As described above, ARGs are often associated with MGEs, which can be transferred across a broad spectrum of bacteria. MGEs include conjugative transposable elements such as transposons and insertion sequences, integrative conjugative elements, or integrons and plasmids (Sorensen et al. 2005; Van Elsas and Bailey 2002). HGT of these elements can increase the spread of ARGs in pathogenic and non-pathogenic bacteria by three principal mechanisms: conjugation that involves plasmid transfer by direct cell-to-cell contact; transformation, which involves the uptake of free DNA via cell wall and integration into the bacterial genome; and transduction by phage-mediated gene transfer (Sorensen et al. 2005).

TWW irrigation can introduce plasmids or other MGEs harboring ARGs to the soil microbiome, and the high mobility and versatility of these elements may contribute to AR propagation far beyond intrinsic elements that are chromosomally harbored in many native soil bacteria. Therefore, when considering the impact of anthropogenic practices on the soil microbiome, it is not only necessary to assess the abundance of ARGs in soil, but also to specifically determine ARG levels that are associated with MGEs, because these elements are the most crucial from a public health perspective. The stability of plasmids and other MGEs in the environment is strongly dictated by a wide array of abiotic parameters such as soil type, micro and macronutrient availability, soil moisture, temperature, O_2, and pH (Rahube and Yost 2010; Van Elsas and Bailey 2002). Future models designed to estimate the effect of TWW irrigation on the propagation of soil AR will need to integrate these factors as well. Furthermore, as presented by Martinez et al. (2007), there are several obstacles that need to be overcome to enable ARG HGT from environmental bacteria to the human-associated microbiota: first, only ARGs that can coexist with human pathogens will contribute to resistance; second, only those genes recruited by gene transfer systems compatible with human pathogens will be transferred; third, elements that produce strong fitness cost in their host will be counter selected; and fourth, when a resistance determinant is acquired by HGT, the probabilities for a new acquisition will be low, unless antibiotic selective pressure changes. However, although many ARGs may disappear from a given habitat due to their high fitness cost, other ARGs may remain in the environment because of their fitness cost is low, and this cost is compensated or even beneficial (Baquero et al. 2009).

The release of TWW effluents into the soil and the balance that governs HGT in bacteria generate important questions associated with the understanding of AR evolution: Is TWW irrigation a significant source of ARGs to native soil bacteria? Can unknown ARGs from native soil bacteria be transferred by HGT to opportunistic and pathogenic bacteria that enter the soil through TWW irrigation? Are the ARGs transferred from environmental bacteria to human pathogens, and if so, do they persistent over time? Do soils irrigated with TWW become ARG hotspots and how frequent does HGT occur in these soils? Although emerging data suggest that this is not the case, the answer to these questions to a large extent is still an enigma and requires additional in-depth analyses.

6.9 CONCLUSIONS

The use of TWW in agriculture is of growing importance, especially in arid and semiarid areas of the world, because it alleviates pressure on natural water sources. However, the treatment of wastewater does not assure the successful removal of all biological and chemical contaminants and differences in quality effluents can be observed among different WWTPs. Thus, biological components and active compounds can be transported from WWTP effluents to terrestrial environments such as plants and soils, impacting in different ways both soil properties and the ecosystems that they support. Certain antibiotics are highly persistent in WWTPs, and ARGs and ARB can be enriched in the wastewater treatment processes, and subsequently, these elements can be transported to the soil by irrigation, where they may be incorporated into the native soil microbiome. Although recent studies seem to indicate that irrigation with TWW does not significantly induce AR reservoirs in soil, the impact of all of the above-mentioned factors is not yet clear especially in the context of mobile ARG transfer between TWW-associated bacteria and the soil. Additional studies are required to answer these questions and to determine the efficient WWTP processes for optimal reduction of ARGs, in order to ensure safe application of TWW for irrigation.

REFERENCES

1. Adrover M, Farrus E, Moya G, Vadell J (2012) Chemical properties and biological activity in soils of Mallorca following twenty years of treated wastewater irrigation. J Environ Manage 95:188–192
2. Aiello A, Marshall B, Levy S, Della-Latta P, Lin S, Larson E (2005) Antibacterial cleaning products and drug resistance. Emerg Infect Dis 11:1565–1570
3. Akiyama T, Asfahl L, Savin C (2010) Broad-host-range plasmids in treated wastewater effluent and receiving streams. J Environ Qual 39:2211–2215
4. Akponikpe P, Wima K, Yakouba H, Mermoud A (2011) Reuse of domestic wastewater treated in macrophyte ponds to irrigate tomato and eggplants in semi-arid West-Africa: benefits and risks. Agr Water Manag 98:834–840
5. Alanis J (2005) Resistance to antibiotics: are we in the post-antibiotic era? Arch Med Res 36:697–705
6. Allen H, Moe L, Rodbumrer J, Gaarder A, Handelsman J (2009) Functional metagenomics reveals diverse β-lactamases in a remote Alaskan soil. ISME J 3:243–251
7. Aminov R, Mackie R (2007) Evolution and ecology of antibiotic resistance genes. FEMS Microbiol Lett 271:147–161
8. An Y, Yoon C, Jung K, Ham J (2007) Estimating the microbial risk of E. coli in reclaimed wastewater irrigation on paddy field. Environ Monit Assess 129:53–60
9. Angin I, Yaganoglu P, Turan M (2005) Effects of long-term wastewater irrigation on soil properties. J Sustain Agr 26:31–42
10. Avisar D, Lester Y, Ronen D (2009) Sulfamethoxazole contamination of a deep phreatic aquifer. Sci Total Environ 407:4278–4282
11. Bahmanyar M (2008) Cadmium, nickel, chromium, and lead levels in soils and vegetables under long-term irrigation with industrial wastewater. Comm Soil Sci Plant Anal 39:2068–2079
12. Baquero F, Alvarez-Ortega C, Martinez J (2009) Ecology and evolution of antibiotic resistance. Environ Microbiol Rep 1:469–476
13. Benveniste R, Davies J (1973) Aminoglycoside antibiotic-inactivating enzymes in Actinomycetes similar to those present in clinical isolates of antibiotic-resistant bacteria. PNAS 70:2276–2280
14. Bernstein N, Guetsky R, Friedman H, Bar-Tal A, Rot I (2008) Monitoring bacterial populations in an agricultural greenhouse production system irrigated with reclaimed wastewater. J Hortic Sci Biotech 83:821–827
15. Bhullar K, Waglechner N, Pawlowski A, Koteva K, Banks E, Johnston M, Barton H, Wright G (2012) Antibiotic resistance is prevalent in an isolated cave microbiome. PLoS One 7:1–11
16. Binh C, Heuer H, Kaupenjohann M, Smalla K (2008) Piggery manure used for soil fertilization is a reservoir for transferable antibiotic resistance plasmids. FEMS Microb Ecol 66:25–37
17. Blumenthal U, Mara D, Peasey A, Ruiz-Palacios G, Stott R (2000) Guidelines for the microbiological quality of treated wastewater used in agriculture: recommendations for revising WHO guidelines. Bull WHO 78:1104–1116

18. Bondarenko S, Gan J, Ernst F, Green R, Baird J, McCullough M (2012) Leaching of pharmaceuticals and personal care products in turfgrass soils during recycled water irrigation. J Environ Qual 41:1268–1274
19. Byrne-Bailey G, Gaze H, Zhang L, Kay P, Boxall A, Hawkey M, Wellington H (2011) Integron prevalence and diversity in manured soil. Appl Environ Microbiol 77:684–687
20. Caccio S, De Giacomo M, Aulicino F, Pozio E (2003) Giardia cystis in wastewater treatment plants in Italy. Appl Environ Microbiol 69:3393–3398
21. Carey C, Lee H, Trevors J (2004) Biology, persistence and detection of Cryptosporidium parvum and Cryptosporidium homynis oocyst. Water Res 38:818–868
22. Chefetz B, Mualem T, Ben-Ari J (2008) Sorption and mobility of pharmaceutical compounds in soil irrigated with reclaimed wastewater. Chemosphere 73:1335–1343
23. Chen Y, Wuang C, Wang Z (2005) Residues and source identification of persistent organic pollutants in farmland soils irrigated by effluents from biological treatment plants. Environ Int 31:778–783
24. Chen F, Ying G, Kong L, Wang L, Zhao J, Zhou L, Zhang L (2011) Distribution and accumulation of endocrine-disrupting chemicals and pharmaceuticals in wastewater irrigated soils in Hebei, China. Environ Pollut 159:1490–1498
25. Ciusa M, Furi L, Knight D, Decorosi F, Fondi M, Raggi C, Coelho J, Aragones L, Moce L, Visa P, Freitas A, Baldassarri L, Fani R, Viti C, Orefici G, Martinez J, Morrissey I, Oggioni M (2012) A novel resistance mechanism to triclosan that suggest horizontal gene transfer and demonstrates a potential selective pressure for reduced biocide susceptibility in clinical strains of Staphylococcus aureus. Int J Antimicrob Ag 40:210–220
26. Czekalski N, Berthold T, Caucci S, Egli A, Burgmann H (2012) Increased levels of multiresistant bacteria and resistance genes after wastewater treatment and their dissemination into Lake Geneva, Switzerland. Front Microbio 3:1–18
27. D'Costa V, McGrann K, Hughes D, Wright G (2006) Sampling the antibiotic resistome. Science 311:374–377
28. D'Costa V, King C, Kalan L, Morar M, Sung W, Schwarz C, Froese D, Zazula G, Calmels F, Debruyne R, Golding B, Poinar H, Wright G (2011) Antibiotic resistance is ancient. Nature 477:457–461
29. DaCosta P, Vaz-Pires P, Bernardo F (2006) Antimicrobial resistance in Enterococcus spp. isolate in inflow, effluent and sludge from municipal sewage water treatment plants. Water Res 40:1735–1740
30. Davies J, Davies D (2010) Origins and evolution of antibiotic resistance. Microbiol Mol Biol R 74:417–433
31. Dzidic S, Suskovic J, Kos B (2008) Antibiotic resistance mechanisms in bacteria: biochemical and genetic aspects. Food Technol Biotechnol 46:11–21
32. Elifanz H, Kautsky L, Mor-Yosef M, Tarchitzky J, Bar-Tal A, Chen Y, Minz D (2011) Microbial activity and organic matter dynamics during 4 years of irrigation with treated wastewater. Microb Ecol 62:973–981
33. EPA (2012) Pharmaceutical and personal care products (PPCPs). EPA web. http://www.epa.gov/ppcp/. Accessed 10 Nov 2012

34. Fatta-Kassinos D, Hapeshi E, Achilleos A, Meric S, Gros M, Petrovic M, Barcelo D (2011a) Existence of pharmaceutical compounds in tertiary-treated urban wastewater that is utilized for reuse applications. Water Resour Manag 25:1183–1193

35. Fatta-Kassinos D, Kalavrouziotis I, Koukoulakis P, Vasquez M (2011b) The risk associated with wastewater reuse and xenobiotics in the agroecological environment. Sci Total Environ 409:3555–3563

36. Fenet H, Mathieu O, Mahjoub O, Li Z, Hillaire-Buyz D, Casellas C, Gomez E (2012) Carbamazepine, carbamazepine epoxide, and dihydroxycarbamazepine sorption to soil and occurrence in a wastewater reuse site in Tunisia. Chemosphere 88:49–54

37. Filip Z, Kanazawa S, Berthelin J (1999) Characterization of effects of a long-term wastewater irrigation on soil quality by microbiological and biochemical parameters. J Plant Nutr Soil Sc 162:409–413

38. Filip Z, Kanazawa S, Berthelin J (2000) Distribution of microorganisms, biomass ATP, and enzymes activities in organic and mineral particles of a long-term wastewater irrigated soil. J Plant Nutr Soil Sc 163:143–150

39. Forsberg K, Reyes A, Wang B, Selleck E, Sommer M, Dantas G (2012) The shared antibiotic resistome of soil bacteria and human pathogens. Science 337:1107–1111

40. Gao P, Munir M, Xagoraraki I (2012) Correlation of tetracycline and sulfonamide antibiotics with corresponding resistance genes and resistant bacteria in a conventional municipal wastewater treatment plant. Sci Total Environ 421:173–183

41. Gelsomino A, Badalucco L, Ambrosoli R, Crecchio C, Puglisi E, Meli S (2006) Changes in chemical and biological soil properties as induced by anthropogenic disturbance: a case study of an agricultural soil under recurrent flooding by wastewaters. Soil Biol Biochem 38:2069–2080

42. Ghosh G, Okuda T, Yamashita N, Tanaka H (2009) Occurrence and elimination of antibiotics at four sewage treatment plants in Japan and their effects on bacterial ammonia oxidation. Water Sci Technol 59:779–786

43. Graber E, Ben-Arie O, Wallach R (2006) Effect of simple disturbance on soil water repellency determination in sandy soils. Geoderma 136:11–19

44. Gullberg E, Cao S, Berg O, Ilback C, Sandegren L, Hughes D, Andersson I (2011) Selection of resistant bacteria at very low antibiotic concentrations. PLoS Pathog 7:1–9

45. Gupta S, Satpati S, Nayek S, Garai D (2010) Effect of wastewater irrigation on vegetables in relation to bioaccumulation of heavy metals and biochemical changes. Environ Monit Assess 165:169–177

46. Gwenzi W, Munondo R (2008) Long-term impacts of pasture irrigation with treated sewage effluent on nutrient status of a sandy soil in Zimbabwe. Nutr Cycl Agroecosys 82:197–207

47. Hey G, Ledin A, la Cour JJ, Andersen H (2012) Removal of pharmaceuticals in biologically treated wastewater by chlorine dioxide or peracetic acid. Environ Technol 33:1041–1047

48. Hidri Y, Bouziri L, Maron P, Anane M, Jedidi N, Hassan A, Ranjard L (2010) Soil DNA evidence for altered microbial diversity after long-term application of municipal wastewater. Agron Sustain Dev 30:423–431

49. Inbar Y (2007) New standards for treated wastewater reuse in Israel. In: Zaidi M (ed) Wastewater reuse-risk assessment, decision-making and environmental security, Springer Netherlands, pp 291–296.

50. Jayaraman R (2009) Antibiotic resistance: an overview of mechanisms and a paradigm shift. Curr Sci 96:1475–1484

51. Jelic A, Gros M, Ginebra A, Cespedes-Sanchez R, Ventura F, Petrovic M, Barcelo D (2011) Occurrence partition and removal of pharmaceutical in sewage water and sludge during wastewater treatment. Water Res 45:1165–1176

52. Jueschke E, Marschner B, Tarchitzky J, Chen Y (2008) Effects of treated wastewater irrigation on the dissolved and soil organic carbon in Israeli soils. Water Sci Technol 57:727–733

53. Khan S, Cao Q, Zheng Y, Huang Y, Zhu Y (2008) Health risks of heavy metals in contaminated soils and food crops irrigated with wastewater in Beijing, China. Environ Pollut 152:686–692

54. Kim S, Park S, Lee J, Benham B, Kim H (2007) Modeling and assessing the impact of reclaimed wastewater irrigation on the nutrition loads from an agricultural watershed containing rice paddy fields. J Environ Sci Heal A 42:305–315

55. Kinney C, Furlong E, Werner S, Cahill J (2006) Presence and distribution of wastewater-derived pharmaceuticals in soil irrigated with reclaimed water. Environ Toxicol Chem 25:317–326

56. Kiziloglu F, Tuean M, Sahin U, Angin I, Anapali O, Okuroglu M (2007) Effects of wastewater irrigation on soil and cabbage-plant (Brassica olereacea var. capitate cv. Yavola-1) chemical properties. J Plant Nutr Soil Sc 170:166–172

57. Knapp W, Dolfing J, Ehlert I, Graham W (2010) Evidence of increasing 429 antibiotic resistance gene abundances in archived soils since 1940. Environ Sci Tech 44:580–587

58. Kummerer K, Henninger A (2003) Promoting resistance by the emission of antibiotics from hospitals and households into effluent. Clinic Microbiol Infec 9:1203–1214

59. LaPara T, Burch T, McNamara P, Tan D, Yan M, Eichmiller J (2011) Tertiary-treated municipal wastewater is significant point source of antibiotic resistance genes into Duluth-Superiro Harbor. Environ Sci Technol 45:9543–9549

60. Li B, Zhang T (2011) Mass flow and removal of antibiotics in two municipal wastewater treatment plants. Chemosphere 83:1284–1289

61. Mañas P, Castro E, Heras J De las (2009) Irrigation with treated wastewater: effects on soil, lettuce (Lactuca sativa) crop and dynamics of microorganisms. J Environ Sci Heal A 44:1261–1273

62. Mapanda F, Mangwayana E, Nyamangara J, Giller K (2005) The effect of long-term irrigation using wastewater on heavy metal contents of soils under vegetables in Harare, Zimbabwe. Agr Ecosyst Environ 107:151–165

63. Martinez J (2009a) Environmental pollution by antibiotics and by antibiotic resistance determinants. Environ Pollut 157:2893–2902

64. Martinez J (2009b) The role of natural environments in the evolution of resistance traits in pathogenic bacteria. Proc Biol Sci 276:2521–2530

65. Martinez J, Baquero F, Andersson D (2007) Predicting antibiotic resistance. Nature 5:958–965

66. Martinez S, Perez-Parra J, Suay R (2011) Use of ozone in wastewater treatment to produce water suitable for irrigation. Water Resour Manag 25:2109–2124
67. McLain J, Williams C (2010) Development of antibiotic resistance in bacteria of soils irrigated with reclaimed wastewater. Dissertation, Phoenix convention center, Arizona
68. Meli S, Porto M, Belligno A, Buffo S, Mazzatura A, Scopa A (2002) Influence of irrigation with lagooned urban wastewater on chemical and microbiological soil parameters in a citrus orchard under Mediterranean condition. Sci Total Environ 285:69–77
69. Mohammad M, Mazahreh N (2003) Changes in soil fertility parameters in response to irrigation of forage crops with secondary treated wastewater. Comm Soil Sci Plant Anal 34:181–1294
70. Mokracka J, Koczura R, Kaznowski A (2012) Multiresistant Enterobacteriaceae with class 1 and class 2 integrons in a municipal wastewater treatment plant. Water Res 46:3353–3363
71. Monteiro S, Boxall A (2010) Occurrence and fate of human pharmaceutical in the environment. Rev Environ Contam T 202:53–154
72. Munir M, Xagoraraki I (2011) Levels of antibiotic resistance genes in manure, biosolids and fertilized soils. J Environ Qual 40:248–255
73. Munir M, Wong K, Xagoraraki I (2011) Release of antibiotic resistant bacteria and genes in the effluent of biosolids of five wastewater utilities in Michigan. Water Res 45:681–693
74. Mutengu S, Hoko Z, Makoni F (2007) An assessment of the public health hazard potential of wastewater reuse for crop production. A case of Bulawayo City, Zimbabwe. Phys Chem Earth 32:1195–1203
75. Ndour N, Baudoin E, Guisse A, Seck M, Khouma M, Brauman A (2008) Impact of irrigation water quality in soil nitrifying and total bacterial communities. Biol Fert Soils 44:797–803
76. Negreanu Y, Pasternack Z, Jurkevitch E, Cytryn E (2012) Impact of treated wastewater irrigation on antibiotic resistance in agricultural soils. Environ Science Technol 46:4800–4808
77. Nikaido M, Tonani K, Juliao F, Trevilato T, Takayanagui A, Sanchez S, Domingo J, Segura-Muñoz S (2010) Analysis of bacteria, parasites and heavy metals in lettuce (Lactuca sativa) and rocket salad (Eruca sativa L.) irrigated with treated effluent from a biological treated wastewater plant. Biol Trace Elem Res 34:342–351
78. Nikolaou A, Meric S, Fatta D (2007) Occurrence patterns of pharmaceuticals in water and wastewater environments. Anal Bioanal Chem 387:1225–1234
79. Novo A, Manaia C (2010) Factors influencing antibiotic resistance burden in municipal wastewater treatment plants. Environ Biotech 87:1157–1166
80. Orlofsky M, Gillor O, Bernstein N, Shapiro K, Miller W, Wuertz S (2011) The correlation between fecal indicator bacteria and pathogens in effluent irrigated tomatoes. 11th Worldwide Workshop for Young Environmental Scientists. Arcueil, France
81. Oved T, Shaviv A, Goldrath T, Mandelbaum R, Minz D (2001) Influence of effluent irrigation on community composition and function of ammonia-oxidizing bacteria in soil. Appl Environ Microbiol 67:3426–3433

82. Pedersen J, Yager M, Suffet I (2003) Xenobiotic organic compounds in runoff from fields irrigated with treated wastewater. J Agr Food Chem 51:1360–1372

83. Pedersen J, Soliman M, Suffet I (2005) Human pharmaceuticals, hormones and personal care product ingredients in runoff from agricultural fields irrigated with treated wastewater. J Agr Food Chem 56:1625–1632

84. Peltier E, Vincent J, Finn C, Graham D (2010) Zinc-induced antibiotic resistance in activated sludge bioreactors. Wat Res 44:3829–3836

85. Phung M, Pham T, Castle J, Rodgers J Jr (2011) Application of water quality guidelines and water quantity calculations to decisions for beneficial use of treated water. App Water Sci 1:85–101

86. Postel S (2000) Entering an era of water scarcity: the challenges ahead. Ecol Appl 10:941–948

87. Pruden A, Pei R, Storteboom H, Carlson K (2006) Antibiotic resistance genes as emerging contaminants: studies in northern Colorado. Environ Sci Technol 40:7445–7450

88. Rahube T, Yost C (2010) Antibiotic resistance plasmids in wastewater treatment plants and their possible dissemination into the environment. AJB 9:9183–9190

89. Rai P, Tripathi B (2007) Microbial contamination in vegetables due to irrigation with partially treated municipal wastewater in a tropical city. Int j Environ Heal R 17:389–395

90. Rezapour S, Samadi A (2011) Soil quality response to long-term wastewater irrigation in inceptisols from a semi-arid environment. Nutr Cycl Agroecosys 91:269–280

91. Sacks M, Bernstein N (2011) Utilization of reclaimed wastewater for irrigation of field-grown melons by surface and subsurface drip irrigation. Isr J Plant Sci 59:159–169

92. Samie A, Obi L, Igumbor O, Momba B (2009) Focus on 14 sewage treatment plants in the Mpumalanga province, South Africa in order to gauge the efficiency of wastewater treatment. Afr Jf Microbiol Res 8:3276–3285

93. Shi Y, Gao L, Li W, Liu J, Cai Y (2012) Investigation of fluoroquinolones, sulfonamides and macrolides in long-term wastewater irrigation soil in Tianjin, China. Bull Environ Contam Toxicol 89:857–861

94. Siemens J, Huschek G, Siebe C, Kaupenjohann M (2008) Concentrations and mobility of human pharmaceuticals in the world's largest wastewater irrigation system, Mexico City-Mezquital valley. Wat Res 42:2124–2134

95. Sim W, Lee J, Lee E, Shin S, Hwang S, Oh J (2011) Occurrence and distribution of pharmaceutical in wastewater from households, livestock farms, hospitals and pharmaceutical manufactures. Chemosphere 82:179–186

96. Song Y, Wilke B, Song X, Gong P, Zhou Q, Yang G (2006) Polycyclic aromatics hydrocarbons (PAHs), polychlorinated biphenyls (PCBs) and heavy metals (HMs) as well as their genotoxicity in soil after long-term wastewater irrigation. Chemosphere 65:1859–1868

97. Sorensen S, Bailey M, Hansen L, Kroer N, Wuertz S (2005) Studying plasmid horizontal transfer in situ: a critical review. Nature 3:700–710

98. Sun K, Zhao Y, Gao B, Liu X, Zhang Z, Xing B (2009) Organochlorine pesticides and polybrominated diphenyl ethers in irrigated soils of Beijing, China: levels, inventory and fate. Chemosphere 77:1199–1205

99. Szczepanowski R, Linke B, Krhan I, Gartemann K, Gutzkow T, Eichler W, Puhler A, Schluter A (2009) Detection of 140 clinically relevant antibiotic-resistance genes in the plasmid metagenome of wastewater treatment plant bacteria showing reduced susceptibility to selected antibiotics. Microbiol 155:2306–2319

100. Tal A (2006) Seeking sustainability: Israel's evolving water management strategies. Science 313:1081–1084

101. Tarchitzky J, Lerner O, Shani U, Arye G, Lowengart-Aycicegi A, Brener A, Chen Y (2007) Water distribution pattern in treated wastewater irrigated soils: hydrophobicity effect. Eur J Soil Sci 58:573–588

102. Topp E, Monteiro S, Beck A, Coehlo B, Boxall A, Duenk P, Kleywegt S, Lapen D, Payne M, Sabourin L, Li H, Metcalfe C (2008) Runoff of pharmaceutical and personal care products following application of biosolid to an agricultural field. Sci Total Environ 369:52–59

103. Toze S (2006) Reuse of effluent water—benefits and risks. Agr Water Manage 80:140–159

104. Tree J, Adams M, Lees D (2003) Chlorination of indicator bacteria and viruses in primary sewage effluents. Appl Environ Microbiol 69:2038–2043

105. Truu M, Truu J, Heinsoo K (2009) Changes in soil microbial community under willow coppice: the effect of irrigation with secondary-treated municipal wastewater. Ecol Eng 35:1011–1020

106. Van Elsas J, Bailey M (2002) The ecology of transfer of mobile genetic elements. FEMS Microbiol Ecol 42:187–197

107. Van Meervenne E, Van Coillie E, Kerckhof M, Devlieghere F, Herman L, De Gelder S, Top M, Boon N (2012) Strain-specific transfer of antibiotic resistance from an environmental plasmid to foodborne pathogens. J Biomed Biotechnol. doi:10.1155/2012/834598

108. Verlicchi P, Aukidy M, Zambello E (2012) Occurrence of pharmaceutical compounds in urban wastewater: removal, mass load and environmental risk after a secondary treatment—a review. Sci Total Environ 429:123–155

109. Volkmann H, Schwartz T, Bischof P, Kirchen S, Obst U (2004) Detection of clinically relevant antibiotic-resistance genes in municipal wastewater using real-time PCR (TaqMan). J Microbiol Meth 56:277–286

110. Walker C, Watson J, Williams C (2012) Occurrence of carbamazepine in soils under different lands uses receiving wastewater. J Environ Qual 41:1263–1267

111. Wang Q, Cui Y, Liu X, Dong Y, Christie P (2006) Soil contamination and plant uptake of heavy metals at polluted sites in China. J Environ Sci Heal 38:823–838

112. Watkinson A, Murby E, Constanzo S (2007) Removal of antibiotics in conventional and advanced wastewater treatment: implications for environmental discharge and wastewater recycling. Water Res 41:4164–4176

113. Xi C, Zhang Y, Marrs C, Ye W, Simon C, Foxman B, Nriagu J (2009) Prevalence of antibiotic resistance in drinking water treatment and distribution systems. App Environ Ecol 75:5714–5718

114. Xu W, Zhang G, Li X, Zou S, Li P, Hu Z, Li J (2007) Occurrence and elimination of antibiotics at four sewage treatment plants in the Pearl River Delta (PRD), South China. Water Res 41:4526–4534

115. Xu J, Wu L, Chen W, Jiang P, Chang A (2009a) Pharmaceutical and personal care products (PPCPs) and endocrine disrupting compounds (EDCs) in runoff from a potato field irrigated with treated wastewater in southern California. J Health Sci 55:306–310

116. Xu J, Chen W, Wu L, Green R, Chang A (2009b) Leachability of some emerging contaminants in reclaimed municipal wastewater-irrigated turf grass fields. Environ Toxicol Chem 28:1842–1850

117. Zhang L, Liu Z (1989) A methodological research on environmental impact assessment of sewage irrigation region. Chi Environ Sci 9:298–303

118. Zhang Y, Marrs C, Simon C, Xi C (2009) Wastewater treatment contributes to selective increase of antibiotic resistance among Acinetobacter spp. Sci Total Environ 407:3702–3706

Metagenomic Profiling of Antibiotic Resistance Genes and Mobile Genetic Elements in a Tannery Wastewater Treatment Plant

ZHU WANG, XU-XIANG ZHANG, KAILONG HUANG, YU MIAO, PENG SHI, BO LIU, CHAO LONG, AND AIMIN LI

7.1 INTRODUCTION

About 210,000 tons of antibiotics are produced annually in China, nearly half of which is used in animal agriculture for sickness prevention and production improvement [1,2]. The improper or illegal use of various antibiotics may result in the accumulation of residues in animal tissues including muscle, liver, kidney, skin and hair [3–5]. Leather production may facilitate the transfer of the antibiotic residues and resistant bacteria from animal tissues to the tannery wastewater. In addition, presence of various heavy metals [6] and biocides [7] in tannery wastewater contributes to co-selection of antibiotics and heavy metals in wastewater treatment plants (WWTPs) [8].

Metagenomic Profiling of Antibiotic Resistance Genes and Mobile Genetic Elements in a Tannery Wastewater Treatment Plant. © Wang Z, Zhang X-X, Huang K, Miao Y, Shi P, et al. (2013). PLoS ONE 8(10): e76079. doi:10.1371/journal.pone.0076079. Creative Commons Attribution License.

Current concerns focus on identification of heavy metal and antibiotic resistant bacteria isolated from tannery wastewater [9,10]. Previous studies have investigated the microbial community of activated sludge in tannery WWTPs through 16S rRNA gene amplification and sequencing [11,12]. However, information about abundance and diversity of antibiotic resistance genes (ARGs) in tannery WWTPs is limited. ARGs are often carried on mobile genetic elements (MGEs) including plasmids [13], transposons [14] and integrons [15], facilitating horizontal transfer among bacteria in WWTPs. Public health problems may arise from the ARGs spread and transfer in the environment.

Recently, metagenomic analysis combined with high-throughput sequencing has been considered as a promising culture-independent method of determining diversity and abundance of ARGs in various environments, such as activated sludge [13], drinking water [16], sediment [17] and soil [18]. This method has also shown great advantages on microbial community profiling due to its unprecedented sequencing depth, which has been used to characterize microbial community structure and function in activated sludge [19], buffalo rumen [20] and pipe biofilm [21].

This study aimed to use Illumina high-throughput sequencing to comprehensively investigate the microbial community structure and function of anaerobic and aerobic sludge in a full-scale tannery wastewater treatment plant, with emphasis on the abundance and diversity of ARGs and MGEs in the sludge.

7.2 MATERIALS AND METHODS

7.2.1 SLUDGE SAMPLING

Activated sludge samples were collected from the full-scale tannery WWTP of Boao Leather Industry Co., Ltd. geographically located in Xiangcheng City (Henan Province, China). We would like to state that the company has approved this study which did not involve endangered or protected species. Basically, a biological treatment system preceded by preliminary treatment including homogenization, chemical coagulation and primary settling was applied in this WWTP (Figure S1). The biological treatment system was composed of an up-flow anaerobic sludge reactor (UASB) and an integrated anoxic/oxic (A/O) reactor (Table S1). Anaerobic sludge was sampled from the UASB, and aerobic sludge was sampled from the last aerobic tank of the A/O reactor (Figure S1). The sludge samples were fixed with 50% ethanol (v/v) on site before transporting to laboratory for DNA extraction.

7.2.2 DNA EXTRACTION

For DNA extraction, 4 ml of the anaerobic sludge and 10 ml of the aerobic sludge were separately centrifuged at 4,000 rpm for 10 min. Approximately 200 mg of pellet was recovered for total genomic DNA extraction in duplicate using FastDNA Soil Kit (MP Biomedicals, CA, USA) following the recommended protocol. The concentration and quality of the extracted DNA were determined with microspectrophotometry (NanoDrop® ND-1000, NanoDrop Technologies, Willmington, DE, USA).

7.2.3 ILLUMINA HIGH-THROUGHPUT SEQUENCING AND QUALITY FILTERING

DNA samples (10 µg each) were sent to Beijing Genome Institute (Shenzhen, China) for high-throughput sequencing using Illumina Hiseq2000. A library consisting of about 180-bp DNA fragment sequences was constructed according to the manufacturer's instructions before DNA sequencing. The strategy "Index 101 PE" (Paired End sequencing, 101-bp reads and 8-bp index sequence) was used for the Illumina sequencing, generating nearly equal amount of data for each sample. The metagenomic data were deposited in the publicly available database of MG-RAST (Meta Genome Rapid Annotation using Subsystem Technology) (http://metagenomics.nmpdr.org) under accession numbers 4494863.3 (anaerobic sludge) and 4494888.3 (aerobic sludge).

For quality control, the sequences contaminated by adapter or containing three or more unknown nucleotides were firstly removed using the quality control (QC) pipeline recommended by Beijing Genome Institute

(Shenzhen, China). FASTX toolkit tools implemented in GALAXY [22] was then used to remove low quality sequences to ensure that more than 75% bases of each filtered read possessed Illumina quality greater than 30 (q30 indicating 0.1% sequencing error rate). The sequences containing one or more unknown nucleotides were removed by using a self-written Python script. The replicate sequences were removed by MG-RAST QC pipeline [23]. After the above quality filtering, a total of 9,194,933 and 8,652,320 quality-filtered reads were obtained for subsequent analysis of anaerobic and aerobic sludge metagenomes, respectively (Table S2).

7.2.4 COMBINED TAXONOMIC CLASSIFICATION AND FUNCTION ANALYSIS

The quality-filtered reads were submitted to the MG-RAST (V3.3) for taxonomic classification and function analysis. Taxonomic analysis was conducted based on all the available annotation source databases in MG-RAST [19]. Both the phylogenetic information contained in the non-redundant database and the similarities to the rRNA databases were obtained for phylogenomic reconstruction of each sample. For functional assignments, the metagenomic data of anaerobic and aerobic sludge were annotated against SEED subsystems in MG-RAST at a cutoff of E-value $< 10^{-5}$ [24]. The SEED established by Argonne National Lab (Argonne, USA) provides a suite of open source tools to enable researchers to create, collect, and maintain sets of gene annotations organized by groups of related biological and biochemical functions across many microorganisms [25]. A SEED subsystem is a collection of functional roles that together create a specific biological process or structural complex, which is created and curated by experts who specialize in an area relating to that subsystem [26]. The annotated sequences were sorted into 28 level 1 subsystems of SEED database to provide an overall profile of microbial functions. For the three level 1 subsystems of protein metabolism, stress response, and virulence, disease and defense, we further investigated specific variations of microbial functions at level 2. Additionally, the level 2 subsystem resistance to antibiotics and toxic compounds was further analyzed at level 3.

7.2.5 ARGS AND MGES ANALYSIS

A local database of resistances genes was created by downloading all sequences from Antibiotic Resistance Database (ARDB) (23,137 sequences
of 380 ARGs encoding resistance to 249 antibiotics) [27]. All quality-filtered reads were compared against the collection of ARGs using BLAST
(BLASTx) at a cutoff of E-value <10^{-5}. A read was annotated as an ARG
according to its best BLAST hit if the hit had a sequence similarity of
above 90% over an alignment of at least 25 amino acids [13,16,17]. Local
databases of insertion sequences (ISs) and integron integrase genes were
separately created by downloading ISs sequences from the ISfinder (2,578
sequences, 22 families of insertion sequences) [28] and integrase genes
from the INTEGRALL (1,447 sequences) [29]. A read was identified as an
insertion sequence or integrase gene if the BLAST hit (BLASTn with the
E-value cut-off at 10^{-5}) had a nucleotide sequence identity of above 90%
over an alignment of at least 50 bp [13,17].

7.3 RESULTS AND DISCUSSION

7.3.1 TAXONOMIC ANALYSIS OF MICROBIAL COMMUNITIES

Taxonomic affiliation of both predicted proteins and rRNA genes sequences in the sludge were conducted based on all the available annotation
source databases in MG-RAST. Bacteria were predominant in both sludge
samples, occupying 88.41% and 93.79% of all annotated sequences in the
anaerobic and aerobic sludge, respectively (Figure S2). *Proteobacteria*
(35.95% and 58.36% of annotated reads from the anaerobic and aerobic
sludge, respectively), *Firmicutes* (16.31% and 6.08%, respectively), *Bacteroidetes* (14.53% and 6.36%, respectively) and *Actinobacteria* (6.66% and
8.06%, respectively) were the dominant phyla in the anaerobic and aerobic
sludge (Figure 1). This result is supported by a previous study indicating that Proteobacteria was the most dominant phylum in sewage sludge,
followed by *Bacteroidetes*, *Firmicutes* and *Actinobacteria* [30]. Microarray [31] and DNA cloning [32] have also shown that *Proteobacteria*

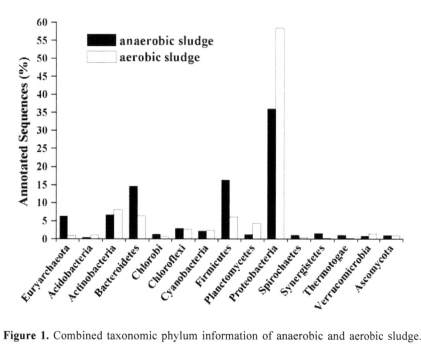

Figure 1. Combined taxonomic phylum information of anaerobic and aerobic sludge. The phyla shown have relative abundance over 1% of total sequencing reads annotated at phylum level in either anaerobic or aerobic sludge.

often dominate in activated sludge. Proteins and carbohydrates often have high concentration in tannery wastewater [33], and *Bacteroidetes* are well known degraders of the organic matters due to the presence of numerous genes encoding protein or carbohydrate degrading enzymes in their genomes [34]. The genomes of *Bacteroidetes* are highly plastic and frequently reorganized, so they can adapt to and dominate in different ecological niches, e.g. soil, ocean, freshwater and activated sludge [30,34].

Oxygen concentration is an important factor shaping microbial community structures in WWTPs, and may make huge contributions to the observed divergence of microbial community structure between anaerobic sludge and aerobic sludge. Our results demonstrated that the phyla of *Synergistetes* and *Thermotogae* (known to be anaerobic bacteria) had higher abundance in the anaerobic sludge than in the aerobic sludge (Figure 1). Lefebvre et al. [11] also indicated that *Synergistetes* occupied 4% of total bacteria population in a UASB treating tannery wastewater, but the phylum was absent in aerobic sludge. A1wR-h+BINL._SL1500_.jpg [35] can use amino acids from the breakdown of proteins and peptides by other organisms, which in return provides short-chain fatty acids and sulfate for terminal degraders including methanogens and sulfate-reducing bacteria. At genus level, aerobic bacteria *Burkholderia* and *Pseudomonas* were predominant in the aerobic sludge (Table S3). The anaerobes *Bacteroides*, *Clostridium* and *Desulfovibrio* dominated in the anaerobic sludge, but they had relatively low abundance in the aerobic sludge (Table S3). As the strictly anaerobic Gram-positive hydrogen–producing bacteria, the genus *Clostridia* was most dominant within the phylum Firmicutes in the anaerobic sludge. This may result from the capability of Clostridia to form endospores to survive under unfavorable environments [36]. It is not surprising that sulfate-reducing bacteria (e.g. *Desulfovibrio*) had high abundance in anaerobic sludge, since sulfate is one of the common pollutants in tannery wastewater [33].

The relative abundance of Archaea in the anaerobic sludge was about three times higher than that of the aerobic sludge. In anaerobic digestors, oxygen unavailability and gentle physical disturbance might contribute to archaeal survival [37,38]. Among Archaea, *Euryarchaeota* had the highest abundance in the anaerobic sludge. Previous studies have confirmed that *Euryarchaeota* dominates in anaerobic sludge by using 16S rRNA gene library analysis [39] and 454-pyrosequencing [40]. This study showed that eukaryotes had nearly equal abundance in the two samples (Figure S2) and the contents of known viruses and other unclassified organisms occupied negligible proportions (<0.28% each) (Figure S2). *Ascomycota*, the largest phylum of Fungi [41], was the most dominant eukaryote in both anaerobic (1.00%) and aerobic sludge (0.99%) (Figure 1).

7.3.2 FUNCTIONAL ANALYSIS OF MICROBIAL COMMUNITIES

Functional analysis was also conducted by using MG-RAST program in the present study. A total of 843,224 (9.17%) sequences of the anaerobic sludge and 600,482 (6.94%) sequences of the aerobic sludge could be annotated against SEED level 1 subsystems database. The annotation proportions were higher than the percentage of successfully assigned sequences (3.03%) reported by Yu and Zhang [19] using Illumina sequencing technology to characterize the structure and function of a sewage sludge community. However, previous studies have reported that about 25% of the Illumina reads and over 36% of the pyrosequencing reads from soil metagenomes had a significant match in the SEED database [42]. Thus, the divergences of annotation proportions may result from the differences in environmental sample types and microbial communities.

Figure 2 shows the relative distribution of 28 basic metabolic categories within the anaerobic and aerobic sludge metagenomes. Protein metabolism was the most abundant category in the anaerobic and aerobic sludge, which is similar to the findings obtained from sewage sludge [19]. However, among the metabolic categories, the category of carbohydrates often has the highest abundance in soil metagenomes [42,43].

Protein metabolism, the most abundant category in both the samples, was selected for further analysis using the MG-RAST program. The annotated sequences of protein metabolism in anaerobic sludge were assigned to five subsystems at level 2, among which protein biosynthesis was the most abundant subsystem (56.87% of annotated sequences in protein metabolism), followed by protein folding (19.01%) and protein degradation (15.58%) (Figure S3A). However, protein degradation had higher abundance than protein folding in aerobic sludge. In aerobic sludge, microorganisms use molecular oxygen (O_2) for respiration or oxidation of nutrients to obtain energy, and inevitably generate reactive oxygen species, such as hydrogen peroxide (H_2O_2) and highly reactive hydroxyl radicals ($\cdot OH$) able to induce oxidative damage to proteins in microorganisms [44]. Microorganisms have to remove oxidized proteins through protein degradation since accumulation of such damaged proteins can cause cellular and organismic dysfunction [44,45]. Additionally, the protein degradation may contribute to energy production in the aerobic sludge where the available

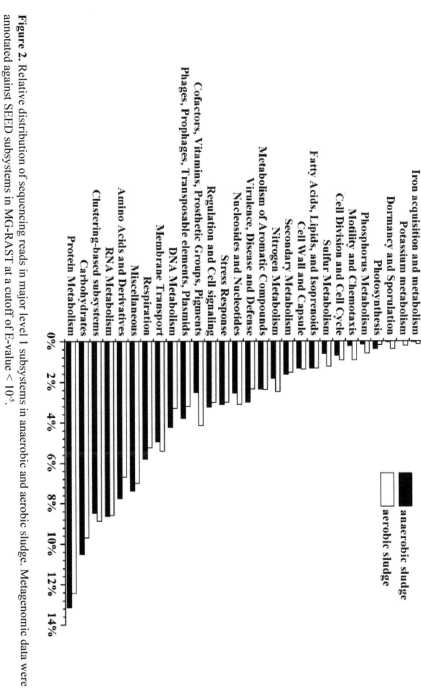

Figure 2. Relative distribution of sequencing reads in major level 1 subsystems in anaerobic and aerobic sludge. Metagenomic data were annotated against SEED subsystems in MG-RAST at a cutoff of E-value < 10⁻⁵.

organic carbon is relatively insufficient in comparison with under anaerobic environment (Table S1) [46].

Figure S3B shows the relative distribution of level 2 subsystems in level 1 category of stress response. Oxidative stress, heat shock, detoxification and osmotic stress were the four most abundant subsystems in anaerobic and aerobic sludge, which might result from the high levels of various toxic chemicals and salts in the extreme environment of tannery wastewater [33]. However, acid stress subsystem was richer in the aerobic sludge than in the anaerobic sludge (Figure S3B). Oxygen availability may facilitate conversion of ammonia to nitrite or nitrate to induce pH decrease [47,48], which might result in the higher level of acid stress in aerobic sludge. Anaerobic sludge had higher abundance of dimethylarginine metabolism subsystems than aerobic sludge (Figure S3B). The aerobic sludge contained high level of nitrate nitrogen (Table S1), and it is known that nitric acid is able to inhibit arginase activity [49].

Figure S3C shows the relative distribution of level 2 subsystems in virulence, disease and defense of anaerobic and aerobic sludge. The genes involved in virulence, disease and defense occupied 3.01% and 2.35% of the total reads annotated by SEED subsystems in the anaerobic and aerobic sludge, respectively. This is generally consistent with abundance of the genes in sewage sludge [19]. Resistance to antibiotics and toxic compounds, an extremely important feature for microbial survival and adaptation in contaminated environments [50], was the most abundant subsystem in both the samples, occupying over 60% of the annotated sequences in the category of virulence, disease and defense in each sample. To better understand antibiotic resistance in the sludge, the subsystem of resistance to antibiotics and toxic compounds was further analyzed at level 3 (Table S4). Both the two samples showed presence of genes conferring resistance to antibiotics (e.g. fluoroquinolones and aminoglycosides) and heavy metals (e.g. copper and arsenic). Fluoroquinolones are widely used as animal feeding additive [51], and fluoroquinolone resistance genes have often been detected in animal breeding farms [52]. Generally, tannery wastewater is characterized with high concentrations of heavy metals [6], and the co-selection of antibiotics and heavy metals may contribute to the prevalence of antibiotic and heavy metal resistance genes in the sludge [8].

7.3.3 ABUNDANCES AND DIVERSITY OF ARGS

In order to comprehensively explore the ARGs present in the tannery WWTPs, we compared all the high-throughput sequencing reads against the ARDB protein database. BLAST analysis showed that a total of 747 reads (0.0081%) of the anaerobic sludge and 877 reads (0.0101%) of the aerobic sludge were assigned to 54 and 42 types of the known ARGs, respectively (Figure 3, Table S5). A total of 48 kinds of multidrug transporters that pump a broad spectrum of antibiotics out of cells were also included in the ARDB database. Due to their contribution to antibiotic resistance phenotype, multidrug transporters have the similar functions of ARGs and are often considered in antibiotic resistance analysis [13,16,17]. In this study, 109 reads (10 types) from anaerobic sludge and 206 reads (12 types) from aerobic sludge were annotated as multidrug transporters.

The proportions of the total ARGs identified in this study were comparable to the results previously obtained from sewage sludge metagenome by Illumina high-throughput sequencing (0.007%) [13] and sewage effluent metagenome by 454-pyrosequencing (0.012%) [53]. However, the annotation proportions of this study were lower than those of antibiotic contaminated sediments (0.02%–1.71%) [17], and higher than those of drinking water (0.0004-0.0071%) [16] and marine water (0.0017%) [53]. Previous studies have shown that sewage treatment plants serve as important reservoirs of environmental ARGs [54,55], and this study reveals that tannery WWTPs can also be considered as the sources of environmental ARGs. The wide use of antibiotics for animal health protection and growth stimulation contributes the prevalence of ARGs in tannery WWTPs [54].

Our results demonstrated that the multidrug resistance genes, tetracycline resistance genes (tet) and sulfonamide resistances genes (sul) were common in the anaerobic sludge, each occupying over 20% of the reads involved in antibiotic resistance (Figure 4). However, in the aerobic sludge, sul genes had the highest abundance (35.46% of the total ARGs reads), followed by the multidrug resistance genes (26.80%) and bacitracin

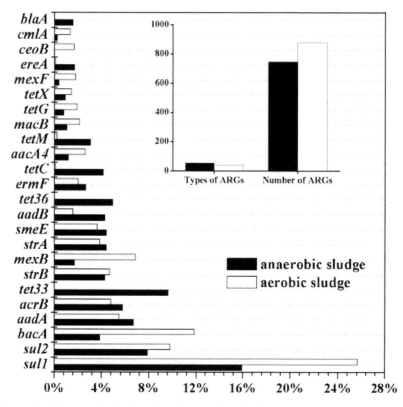

Figure 3. Types and relative abundance of antibiotic resistance genes (ARGs) in anaerobic sludge and aerobic sludge. The ARGs shown have relative abundance over 1% of the total ARGs reads in either anaerobic or aerobic sludge.

resistance genes (11.86%) (Figure 4). The prevalence of tet and sul genes in the two activated sludge samples may result from the frequent use of tetracycline and sulfonamides for livestock purposes in China [56,57].

Among the identified ARGs, the dihydropteroate synthase gene sul1 that confer resistance to sulfonamides had the highest abundance in both the anaerobic and aerobic sludge (Figure 3). Besides sul1, sulfonamide resistance gene sul2 also showed high levels, occupying 7.90% and

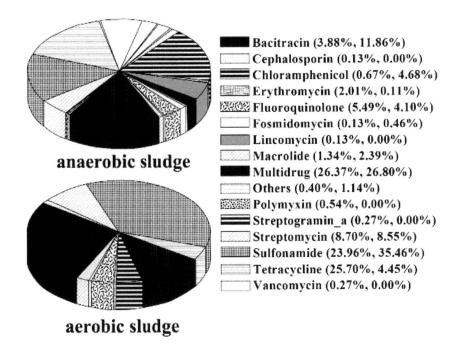

Figure 4. Antibiotic resistance gene patterns in anaerobic and aerobic sludge. The resistance genes were grouped after alignment of the high-throughput sequencing reads against ARDB database. The two percentages shown in the brackets represents the proportions of the reads of each antibiotic resistance gene (ARG) in the total reads of all the identified ARGs in anaerobic (first number) or aerobic (last number) sludge.

9.81% of total ARGs reads in the anaerobic and aerobic sludge, respectively (Figure 3). Sulfonamides with high solubility can persist in the environment [1,58], resulting in the high abundance of sul1 and sul2 in the tannery WWTP. Tet genes were highly rich in the anaerobic sludge, occupying 25.70% of total ARGs reads, but only 4.45% were annotated as tet genes in aerobic sludge (Figure 4). Among the tet genes, tet33 had the highest abundance (72 reads, 9.64% of the total ARGs' reads) in the anaerobic sludge, while the gene was absent in the aerobic sludge. Similarly, tetC, tet36 and tetM were common in the anaerobic sludge, but they had lower abundance or were absent in the aerobic sludge (Figure S4). However, tetC often had higher levels than other tet genes in the aerobic tank of sewage treatment plants [55]. Tetracycline is not biodegradable and can be easily adsorbed to sludge [59]. In this study, the UASB was run under long sludge retention time with high biomass (Table S1), which may facilitate adsorption of tetracycline to sludge, subsequently resulting in higher abundance of the tet genes in the anaerobic sludge than in the aerobic sludge.

It should be noted that the results of ARGs abundance and diversity obtained by ARDB alignment were different from those derived from MG-RAST analysis. The divergence may result from the difference in reference databases and alignment methods. The BLAST program was used for ARDB-based analysis in this study, but MG-RAST relies on BLAT for similarity search, which is less sensitive than BLAST [60]. ARDB unified most of the publicly available ARGs and is considered as a comprehensive and higher-coverage annotation source for ARGs analysis [13,17]. However, the subsystem of "Resistance to antibiotics and toxic compounds" within SEED database contains incomplete information of ARGs (http://theseed.uchicago.edu/FIG/subsys.cgi).

7.3.4 ABUNDANCES AND DIVERSITY OF MGES

The mobility and acquisition of ARGs depends on MGEs, such as plasmids, transposons, ISs and integrons. In this study we focused our analysis on ISs and integrons. Search in INTEGRALL database showed that a total of 130 reads (0.0014%) of the anaerobic sludge and 327 reads (0.0038%) of the aerobic sludge could be assigned to integrase genes (Table S6). The most abundant integrase gene was intI1, which occupied 80.00% and 76.45% of alignment hits of anaerobic sludge and aerobic sludge, respectively. Previous studies have also shown the prevalence of integrons in WWTPs [61], including class 1 integrons carrying various ARGs in both aerobic and anaerobic sewage sludges [54,62]. It was found that 2 sequencing reads from the aerobic sludge could be annotated as IntINeu, a chromosomal integron integrase gene from *Nitrosomonas europaea*, which has been shown to be able to excise and integrate several resistance gene cassettes [63]. This is consistent with our results that the genera Nitrosomonas had higher abundance in aerobic sludge than in anaerobic sludge (Table S3).

Alignment against the ISfinder database demonstrated that a total of 586 reads (0.0064%) of the anaerobic sludge and 687 reads (0.0079%) of the aerobic sludge could match 76 and 81 types of known ISs, respectively. However, the two samples shared only 29 common types (Table S7). Among the ISs in the anaerobic sludge, ISEfa4 (133 reads, 22.70%) had the highest abundance, followed by ISEcp1 (129 reads, 22.01%) and ISDde1 (99 reads, 16.89%), but they had much lower abundance or were absent in the aerobic sludge. ISDde1 is usually located in the cells of strictly anaerobic sulfate-reducing bacteria *Desulfovibrio desulfuricans* [64], and ISEfa4 is often carried by gut pathogen *Enterococcus faecium* [65]. This is consistent with our results that the genera *Desulfovibrio* and *Enterococcus* had higher abundance in anaerobic sludge than in aerobic sludge (Table S3). Different from the anaerobic sludge, the aerobic sludge was dominated by ISVsa3 (106 reads, 15.43%), ISSm2 (70 reads, 10.19%) and ISPps1 (49 reads, 7.13%) (Table S7). ISPps1 prevalent in activated sludge [13] and drinking water [16] has a very broad host range including Gram-negative (*Alpha-*, *Beta-*, and *Gamma-Proteobacteria*) and Gram-positive bacteria (*Arthrobacter aurescens* TC1) [66].

Our results suggested that integrons and ISs prevalent in the two sludge samples might play important roles in acquisition and mobility of various ARGs among the bacterial species. Therefore, the discharge of the tannery wastewater into the environments may be of great public health concern, since the MGEs in surface water and groundwater could potentially transfer antibiotic resistance to the bacteria in drinking water or food chain [67].

In conclusion, this study demonstrated that high-throughput sequencing provided a comprehensive insight in microbial community structures and functions of the aerobic and anaerobic sludge in tannery WWTPs. Metagenomic analysis revealed prevalence of a variety of ARGs in tannery WWTPs. Sulfonamide resistance genes had high abundance in both the sludge samples, but tetracycline resistance genes preferred anaerobic environment. Various MGEs including integrons and ISs were prevalent in the tannery WWTP.

REFERENCES

1. Luo Y, Xu L, Rysz M, Wang Y, Zhang H et al. (2011) Occurrence and transport of tetracycline, sulfonamide, quinolone, and macrolide antibiotics in the Haihe River Basin, China. Environ Sci Technol 45: 1827-1833. doi:10.1021/es104009s. PubMed: 21309601.
2. Hvistendahl M (2012) Public health. China takes aim at rampant antibiotic resistance. Science 336: 795. doi:10.1126/science.336.6083.795. PubMed: 22605727.
3. Goto T, Ito Y, Yamada S, Matsumoto H, Oka H (2005) High-throughput analysis of tetracycline and penicillin antibiotics in animal tissues using electrospray tandem mass spectrometry with selected reaction monitoring transition. J Chromatogr A 1100: 193-199. doi:10.1016/j.chroma.2005.09.056. PubMed: 16214156.
4. Gratacos-Cubarsi M, Castellari M, Garcia-Regueiro JA (2006) Detection of sulphamethazine residues in cattle and pig hair by HPLC-DAD. J Chromatogr B Anal Technol Biomed Life Sci 832: 121-126. doi:10.1016/j.jchromb.2006.01.002.
5. Bittencourt MS, Martins MT, de Albuquerque FG, Barreto F, Hoff R (2012) High-throughput multiclass screening method for antibiotic residue analysis in meat using liquid chromatography-tandem mass spectrometry: a novel minimum sample preparation procedure. Foods Addit Contam A Chem Anal Control Expo Risk Assess 29: 508-516. doi:10.1080/19440049.2011.606228. PubMed: 21988236.
6. Tariq SR, Shah MH, Shaheen N, Khalique A, Manzoor S et al. (2006) Multivariate analysis of trace metal levels in tannery effluents in relation to soil and water: a case study from Peshawar, Pakistan. J Environ Manage 79: 20-29. doi:10.1016/j.jenvman.2005.05.009. PubMed: 16154685.
7. Tisler T, Zagorc-Koncan J, Cotman M, Drolc A (2004) Toxicity potential of disinfection agent in tannery wastewater. Water Res 38: 3503-3510. doi:10.1016/j.watres.2004.05.011. PubMed: 15325176.
8. Baker-Austin C, Wright MS, Stepanauskas R, McArthur JV (2006) Co-selection of antibiotic and metal resistance. Trends Microbiol 14: 176-182. doi:10.1016/j.tim.2006.02.006. PubMed: 16537105.
9. Verma T, Srinath T, Gadpayle RU, Ramteke PW, Hans RK et al. (2001) Chromate tolerant bacteria isolated from tannery effluent. Bioresoure Technol 78: 31-35. doi:10.1016/S0960-8524(00)00168-1. PubMed: 11265785.
10. Alam MZ, Ahmad S, Malik A (2011) Prevalence of heavy metal resistance in bacteria isolated from tannery effluents and affected soil. Environ Monit Assess 178: 281-291. doi:10.1007/s10661-010-1689-8. PubMed: 20824329.
11. Lefebvre O, Vasudevan N, Thanasekaran K, Moletta R, Godon JJ (2006) Microbial diversity in hypersaline wastewater: the example of tanneries. Extremophiles 10: 505-513. doi:10.1007/s00792-006-0524-1. PubMed: 16738814.
12. Chen J, Tang Y-Q, Wu X-L (2012) Bacterial community shift in two sectors of a tannery plant and its Cr (VI) removing potential. Geomicrobiol J 29: 226-235. doi:10.1080/01490451.2011.558562.
13. Zhang T, Zhang XX, Ye L (2011) Plasmid metagenome reveals high levels of antibiotic resistance genes and mobile genetic elements in activated sludge. PLOS ONE 6: e26041. doi:10.1371/journal.pone.0026041. PubMed: 22016806.

14. Tennstedt T, Szczepanowski R, Krahn I, Pühler A, Schlüter A (2005) Sequence of the 68,869 bp IncP-1α plasmid pTB11 from a waste-water treatment plant reveals a highly conserved backbone, a Tn402-like integron and other transposable elements. Plasmid 53: 218-238. doi:10.1016/j.plasmid.2004.09.004. PubMed: 15848226.

15. Ma L, Zhang XX, Cheng S, Zhang Z, Shi P et al. (2011) Occurrence, abundance and elimination of class 1 integrons in one municipal sewage treatment plant. Ecotoxicology 20: 968-973. doi:10.1007/s10646-011-0652-y. PubMed: 21431316.

16. Shi P, Jia S, Zhang XX, Zhang T, Cheng S et al. (2013) Metagenomic insights into chlorination effects on microbial antibiotic resistance in drinking water. Water Res 47: 111-120. doi:10.1016/j.watres.2012.09.046. PubMed: 23084468.

17. Kristiansson E, Fick J, Janzon A, Grabic R, Rutgersson C et al. (2011) Pyrosequencing of antibiotic-contaminated river sediments reveals high levels of resistance and gene transfer elements. PLOS ONE 6: e17038. doi:10.1371/journal.pone.0017038. PubMed: 21359229.

18. Monier JM, Demanèche S, Delmont TO, Mathieu A, Vogel TM et al. (2011) Metagenomic exploration of antibiotic resistance in soil. Curr Opin Microbiol 14: 229-235. doi:10.1016/j.mib.2011.04.010. PubMed: 21601510.

19. Yu K, Zhang T (2012) Metagenomic and metatranscriptomic analysis of microbial community structure and gene expression of activated sludge. PLOS ONE 7: e38183. doi:10.1371/journal.pone.0038183. PubMed: 22666477.

20. Singh KM, Jakhesara SJ, Koringa PG, Rank DN, Joshi CG (2012) Metagenomic analysis of virulence-associated and antibiotic resistance genes of microbes in rumen of Indian buffalo (Bubalus bubalis). Gene 507: 146-151. doi:10.1016/j.gene.2012.07.037. PubMed: 22850272.

21. Gomez-Alvarez V, Revetta RP, Santo Domingo JW (2012) Metagenome analyses of corroded concrete wastewater pipe biofilms reveal a complex microbial system. BMC Microbiol 12: 122. doi:10.1186/1471-2180-12-122. PubMed: 22727216.

22. Goecks J, Nekrutenko A, Taylor J, Galaxy Team (2010) Galaxy: a comprehensive approach for supporting accessible, reproducible, and transparent computational research in the life sciences. Genome Biol 11: R86. doi:10.1186/gb-2010-11-8-r86. PubMed: 20738864.

23. Gomez-Alvarez V, Teal TK, Schmidt TM (2009) Systematic artifacts in metagenomes from complex microbial communities. ISME J, 3: 1314-1317. PubMed: 19587772.

24. Pfister CA, Meyer F, Antonopoulos DA (2010) Metagenomic profiling of a microbial assemblage associated with the California mussel: a node in networks of carbon and nitrogen cycling. PLOS ONE 5: e10518. doi:10.1371/journal.pone.0010518. PubMed: 20463896.

25. Overbeek R, Disz T, Stevens R (2004) A Peer-to-Peer Environment for Annotation of Genomes: The SEED.

26. Overbeek R, Begley T, Butler RM, Choudhuri JV, Chuang HY et al. (2005) The subsystems approach to genome annotation and its use in the project to annotate 1000 genomes. Nucleic Acids Res 33: 5691-5702. doi:10.1093/nar/gki866. PubMed: 16214803.

27. Liu B, Pop M (2009) ARDB--Antibiotic Resistance Genes Database. Nucleic Acids Res 37: D443-D447. doi:10.1093/nar/gkn656. PubMed: 18832362.

28. Siguier P, Perochon J, Lestrade L, Mahillon J, Chandler M (2006) ISfinder: the reference centre for bacterial insertion sequences. Nucleic Acids Res 34: D32-D36. doi:10.1093/nar/gkj014. PubMed: 16381877.

29. Moura A, Soares M, Pereira C, Leitão N, Henriques I et al. (2009) INTEGRALL: a database and search engine for integrons, integrases and gene cassettes. Bioinformatics 25: 1096-1098. doi:10.1093/bioinformatics/btp105. PubMed: 19228805.

30. Zhang T, Shao MF, Ye L (2011) 454 Pyrosequencing reveals bacterial diversity of activated sludge from 14 sewage treatment plants. ISME J 6: 1137-1147. PubMed: 22170428.

31. Xia S, Duan L, Song Y, Li J, Piceno YM et al. (2010) Bacterial community structure in geographically distributed biological wastewater treatment reactors. Environ Sci Technol 44: 7391-7396. doi:10.1021/es101554m. PubMed: 20812670.

32. Snaidr J, Amann R, Huber I, Ludwig W, Schleifer KH (1997) Phylogenetic analysis and in situ identification of bacteria in activated sludge. Appl Environ Microbiol 63: 2884-2896. PubMed: 9212435.

33. Murugananthan M, Bhaskar Raju G, Prabhakar S (2004) Separation of pollutants from tannery effluents by elector flotation. Sep Purif Technol 40: 69-75. doi:10.1016/j.seppur.2004.01.005.

34. Thomas F, Hehemann JH, Rebuffet E, Czjzek M, Michel G (2011) Environmental and gut bacteroidetes: the food connection. Front Microbiol 2: 93. PubMed: 21747801.

35. Vartoukian SR, Palmer RM, Wade WG (2007) The division "Synergistes". Anaerobe 13: 99-106. doi:10.1016/j.anaerobe.2007.05.004. PubMed: 17631395.

36. Cheong DY, Hansen CL (2006) Bacterial stress enrichment enhances anaerobic hydrogen production in cattle manure sludge. Appl Microbiol Biotechnol 72: 635-643. doi:10.1007/s00253-006-0313-x. PubMed: 16525779.

37. Godon JJ, Zumstein E, Dabert P, Habouzit F, Moletta R (1997) Molecular microbial diversity of an anaerobic digester as determined by small-subunit rDNA sequence analysis. Appl Environ Microbiol 63: 2802-2813. PubMed: 9212428.

38. Santegoeds CM, Damgaard LR, Hesselink G, Zopfi J, Lens P et al. (1999) Distribution of sulfate-reducing and methanogenic bacteria in anaerobic aggregates determined by microsensor and molecular analyses. Appl Environ Microbiol 65: 4618-4629. PubMed: 10508098.

39. Pender S, Toomey M, Carton M, Eardly D, Patching JW et al. (2004) Long-term effects of operating temperature and sulphate addition on the methanogenic community structure of anaerobic hybrid reactors. Water Res 38: 619-630. doi:10.1016/j.watres.2003.10.055. PubMed: 14723931.

40. Kröber M, Bekel T, Diaz NN, Goesmann A, Jaenicke S et al. (2009) Phylogenetic characterization of a biogas plant microbial community integrating clone library 16S-rDNA sequences and metagenome sequence data obtained by 454-pyrosequencing. J Biotechnol 142: 38-49. doi:10.1016/j.jbiotec.2009.02.010. PubMed: 19480946.

41. James TY, Kauff F, Schoch CL, Matheny PB, Hofstetter V et al. (2006) Reconstructing the early evolution of Fungi using a six-gene phylogeny. Nature 443: 818-822. doi:10.1038/nature05110. PubMed: 17051209.

42. Uroz S, Ioannidis P, Lengelle J, Cébron A, Morin E et al. (2013) Functional assays and metagenomic analyses reveals differences between the microbial communities inhabiting the soil horizons of a Norway spruce plantation. PLOS ONE 8: e55929. doi:10.1371/journal.pone.0055929. PubMed: 23418476.

43. Urich T, Lanzén A, Qi J, Huson DH, Schleper C et al. (2008) Simultaneous assessment of soil microbial community structure and function through analysis of the meta-transcriptome. PLOS ONE 3: e2527. doi:10.1371/journal.pone.0002527. PubMed: 18575584.

44. Cabiscol E, Tamarit J, Ros J (2000) Oxidative stress in bacteria and protein damage by reactive oxygen species. Int Microbiol 3: 3-8. PubMed: 10963327.

45. Goldberg AL (2003) Protein degradation and protection against misfolded or damaged proteins. Nature 426: 895-899. doi:10.1038/nature02263. PubMed: 14685250.

46. Reeve CA, Bockman AT, Matin A (1984) Role of protein degradation in the survival of carbon-starved Escherichia coli and Salmonella typhimurium. J Bacteriol 157: 758-763. PubMed: 6365890.

47. Marcelino M, Wallaert D, Guisasola A, Baeza JA (2011) A two-sludge system for simultaneous biological C, N and P removal via the nitrite pathway. Water Sci Technol 64: 1142-1147. doi:10.2166/wst.2011.398. PubMed: 22214063.

48. Vejmelkova D, Sorokin DY, Abbas B, Kovaleva OL, Kleerebezem R et al. (2012) Analysis of ammonia-oxidizing bacteria dominating in lab-scale bioreactors with high ammonium bicarbonate loading. Appl Microbiol Biotechnol 93: 401-410. doi:10.1007/s00253-011-3409-x. PubMed: 21691786.

49. Morris SM (2007) Arginine metabolism: boundaries of our knowledge. J Nutr 137: 1602S-1609S. PubMed: 17513435.

50. Silva CC, Hayden H, Sawbridge T, Mele P, Kruger RH et al. (2012) Phylogenetic and functional diversity of metagenomic libraries of phenol degrading sludge from petroleum refinery wastewater treatment system. AMB Express 2: 18. doi:10.1186/2191-0855-2-18. PubMed: 22452812.

51. Borràs S, Ríos-Kristjánsson JG, Companyó R, Prat MD (2012) Analysis of fluoroquinolones in animal feeds by liquid chromatography with fluorescence detection. J Sep Sci 35: 2048-2053. doi:10.1002/jssc.201200302. PubMed: 22778021.

52. Li J, Wang T, Shao B, Shen J, Wang S et al. (2012) Plasmid-mediated quinolone resistance genes and antibiotic residues in wastewater and soil adjacent to swine feedlots: potential transfer to agricultural lands. Environ Health Perspect 120: 1144-1149. doi:10.1289/ehp.1104776. PubMed: 22569244.

53. Port JA, Wallace JC, Griffith WC, Faustman EM (2012) Metagenomic profiling of microbial composition and antibiotic resistance determinants in Puget Sound. PLOS ONE 7: e48000. doi:10.1371/journal.pone.0048000. PubMed: 23144718.

54. Zhang XX, Zhang T, Fang HH (2009) Antibiotic resistance genes in water environment. Appl Microbiol Biotechnol 82: 397-414. doi:10.1007/s00253-008-1829-z. PubMed: 19130050.

55. Zhang XX, Zhang T (2011) Occurrence, abundance, and diversity of tetracycline resistance genes in 15 sewage treatment plants across China and other global locations. Environ Sci Technol 45: 2598-2604. doi:10.1021/es103672x. PubMed: 21388174.

56. Xu W, Zhang G, Li X, Zou S, Li P et al. (2007) Occurrence and elimination of antibiotics at four sewage treatment plants in the Pearl River Delta (PRD), South China. Water Res 41: 4526-4534. doi:10.1016/j.watres.2007.06.023. PubMed: 17631935.

57. Zhao L, Dong YH, Wang H (2010) Residues of veterinary antibiotics in manures from feedlot livestock in eight provinces of China. Sci Total Environ 408: 1069-1075. doi:10.1016/j.scitotenv.2009.11.014. PubMed: 19954821.

58. Heise J, Höltge S, Schrader S, Kreuzig R (2006) Chemical and biological characterization of non-extractable sulfonamide residues in soil. Chemosphere 65: 2352-2357. doi:10.1016/j.chemosphere.2006.04.084. PubMed: 16774778.

59. Kim S, Eichhorn P, Jensen JN, Weber AS, Aga DS (2005) Removal of antibiotics in wastewater: Effect of hydraulic and solid retention times on the fate of tetracycline in the activated sludge process. Environ Sci Technol 39: 5816-5823. doi:10.1021/es050006u. PubMed: 16124320.

60. Yu K, Zhang T (2013) Construction of Customized Sub-Databases from NCBI-nr Database for Rapid Annotation of Huge Metagenomic Datasets Using a Combined BLAST and MEGAN Approach. PLOS ONE 8: e59831. doi:10.1371/journal.pone.0059831. PubMed: 23573212.

61. Moura A, Henriques I, Smalla K, Correia A (2010) Wastewater bacterial communities bring together broad-host range plasmids, integrons and a wide diversity of uncharacterized gene cassettes. Res Microbiol 161: 58-66. doi:10.1016/j.resmic.2009.11.004. PubMed: 20004718.

62. Zhang XX, Zhang T, Zhang M, Fang HH, Cheng SP (2009) Characterization and quantification of class 1 integrons and associated gene cassettes in sewage treatment plants. Appl Microbiol Biotechnol 82: 1169-1177. doi:10.1007/s00253-009-1886-y. PubMed: 19224208.

63. Léon G, Roy PH (2003) Excision and integration of cassettes by an integron integrase of Nitrosomonas europaea. J Bacteriol 185: 2036-2041. doi:10.1128/JB.185.6.2036-2041.2003. PubMed: 12618471.

64. Hauser LJ, Land ML, Brown SD, Larimer F, Keller KL et al. (2011) Complete genome sequence and updated annotation of Desulfovibrio alaskensis G20. J Bacteriol 193: 4268-4269. doi:10.1128/JB.05400-11. PubMed: 21685289.

65. Depardieu F, Reynolds PE, Courvalin P (2003) VanD-type vancomycin-resistant Enterococcus faecium 10/96A. Antimicrob Agents Chemother 47: 7-18. doi:10.1128/AAC.47.1.7-18.2003. PubMed: 12499162.

66. Schleinitz KM, Vallaeys T, Kleinsteuber S (2010) Structural characterization of ISCR8, ISCR22, and ISCR23, subgroups of IS91-like insertion elements. Antimicrob Agents Chemother 54: 4321-4328. doi:10.1128/AAC.00006-10. PubMed: 20625149.

67. Chee-Sanford JC, Aminov RI, Krapac IJ, Garrigues-Jeanjean N, Mackie RI (2001) Occurrence and diversity of tetracycline resistance genes in lagoons and groundwater underlying two swine production facilities. Appl Environ Microbiol 67: 1494-1502. doi:10.1128/AEM.67.4.1494-1502.2001. PubMed: 11282596.

There are several supplemental files that are not available in this version of the article. To view this additional information, please use the citation on the first page of this chapter.

CHAPTER 8

A Comprehensive Insight into Tetracycline Resistant Bacteria and Antibiotic Resistance Genes in Activated Sludge Using Next-Generation Sequencing

KAILONG HUANG, JUNYING TANG, XU-XIANG ZHANG, KE XU, AND HONGQIANG REN

8.1 INTRODUCTION

Extensive use and abuse of antibiotics in health protection and agricultural production have led to the emergence of widespread various antibiotic resistance genes (ARGs) and resistant bacteria (ARB) in the environment [1,2], which is thought to pose an ever increasing threat to public health [3]. The broad spectrum tetracyclines are one of the most frequently used classes of antibiotics for protection of human and animal health [4]. Previous studies have shown that the concentrations of tetracycline in livestock wastewater [5,6] and municipal sewage [7,8] were 4.1~32.67 µg/L and 89.4~652.6 ng/L, respectively. Increasing evidence suggested that sewage treatment plants (STPs) serve as important reservoirs for environmental tetracycline resistant bacteria (TRB) and resistance genes (tet) [9,10,11].

Both culture-based [9,10,12] and culture-independent approaches [13] have been used to explore the TRB in STPs. Classical microbiological methodology relies on plate counting of coliforms, which makes the assessment results unrepresentative and biased. Currently, molecular methods used for exploring TRB in sludge include polymerase chain reaction-denaturing gradient gel electrophoresis (PCR-DGGE) [14], quantitative real time PCR (qPCR) [14], molecular cloning [11] and microarray [15], but the methods are time- and cost-consuming due to low throughput. Recently, growing evidence has shown that next-generation sequencing is a powerful metagenomic tool for comprehensive overview of microbial communities and/or functional genes in various environmental compartments, including soil [16], human gastrointestinal tract [17], sediments [18], and wastewater treatment plants [19].

In this study, we designed a batch experiment to culture STP sludge in filtered sewage fed with different concentrations of tetracycline to identify TRB community composition in the sludge and to evaluate the effect of tetracycline stress on the abundance and diversity of tet genes. 454 Pyrosequencing was used to explore the TRB in activated sludge based on PCR of bacterial 16S rRNA gene. Illumina high-throughput sequencing in combination with qPCR and molecular cloning were also employed to investigate the relative abundance and diversity of ARGs including tet genes. This study revealed the distribution patterns of TRB and ARGs in activated sludge and provided a useful tool for comprehensive investigation of tetracycline resistance in the environment.

8.2 RESULTS AND DISCUSSION

8.2.1 BACTERIAL COMMUNITY SHIFT UNDER TETRACYCLINE STRESS

Pyrosequencing of 16S rRNA gene showed that 6-day tetracycline treatment separately at 1, 5 and 20 mg/L tended to increase the number of operational taxonomic units (OTUs) in the sludge, which agrees with the patterns of Chao 1 and Shannon index (Table 1). The reason may be that the growth of the dominant species in the sludge was inhibited under tetracycline stress, while more species with low abundance had the opportunity to survive and reproduce to reach the detection limit. Li et al. [20] also indicated that the antibiotic stresses seemed not effective in reduction of the bacterial diversities of river water. Interestingly, the sludge fed with 5 mg/L had the richest diversity (1692 OTUs), and 1 mg/L tetracycline treatment also increased the OTUs number, revealing that subinhibitory concentrations of tetracycline stress may favor enhancement of species richness [21].

As shown in Figure 1, Acidobacteria (27.3%) was the most abundant phylum in the sludge without tetracycline treatment, followed by *Proteobacteria* (11.6%), Actinobacteria (11.2%), Planctomycetes (5.9%), *Chloroflexi* (5.3%), Bacteroidetes (1.8%), TM 7 (1.7%), WS3 (1.1%), Nitrospira (0.7%) and Firmicutes (0.5%). Lozada et al. [22] also indicated that *Proteobacteria* and Acidobacteria were dominant in surfactant-enrichment lab-scale activated sludge. Acidobacteria, a common and predominant phylum in sludge [22], seems susceptible to tetracycline since the phylum abundance decreased from 27.3% under no tetracycline stress to 6.2% with 20 mg/L tetracycline treatment. Actinobacteria and Planctomycetes were also susceptible to tetracycline since their abundance

TABLE 1. Number of 16S rRNA gene sequences analyzed, observed OTUs, Chao 1 and Shannon index for each sample at similarity of 97%.

Tetracycline Concentrations	No. of Raw Sequences	Observed OTUs	Chao 1	Shannon Index
0 mg/L	7097	1112	1562	5.94
1 mg/L	13,351	1306	1988	6.19
5 mg/L	9306	1692	2899	6.60
20 mg/L	12,802	1347	1975	6.35

OTUs: Operational taxonomic units; Chao 1: Chao 1 estimator.

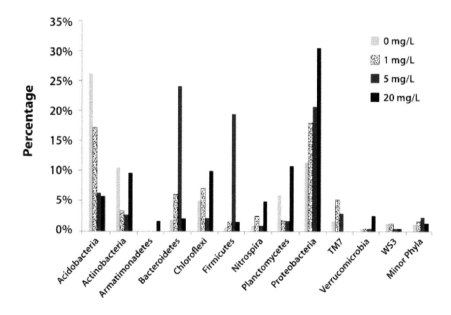

Figure 1. Abundance of various bacterial phyla in sludge after 6 days treatment with different concentrations of tetracycline (0~20 mg/L). The filtered pyrosequencing reads were classified using RDP Classifier at a confidence threshold of 80%. The relative abundance is presented as the percentage of each phylum in total effective reads of the corresponding sample.

evidently decreased with the increase of tetracycline concentration. On the contrary, tetracycline treatment dramatically increased the abundance of *Proteobacteria* in the sludge. Bacteriodetes and Firmicutes seemed to have higher abundance after 5 mg/L tetracycline treatment, but had lower abundance after 20 mg/L tetracycline treatment.

Figure 2 shows that at the level of genus, *Gp16* (14.9%) dominated in the sludge without tetracycline treatment, followed by *Gp17* (5.5%), *Gp6* (5.4%), *Caldilinea* (2.5%), *Singulisphaera* (2.1%), *TM7_genera_ incertae_sedis* (1.7%), *Sphaerobacter* (1.1%), *WS3_genera_incertae_ sedis* (1.1%) and *Conexibacter* (1.0%). Culture with 5 mg/L tetracycline decreased the abundance of *Gp16, Gp17, Gp6, Singulisphaera, Conexibacter* and *TM7_genera_incertae_sedis*. It should be noted that tetracycline treatment at subinhibitory concentrations (5 mg/L) considerably reduced *Acidobacteria* abundance. Within the *Acidobacteria* phylum, the *Gp16* genus was found very susceptible to tetracycline, since its abundance was 14.9%, 5.7% and 2.1% with tetracycline at 0, 1, and 5 mg/L, respectively (Figure 2). The subinhibitory-dose treatment tended to increase the abundance of *Bacteroidetes* and *Firmicutes* phyla (Figure 1), as well as *Azonexus, Methyloversatilis* and *Perlucidibaca* genera (Figure 2). Various bacterial strains of *Bacteroidetes, Firmicutes* and *Proteobacteria* have previously been isolated from livestock feces, farmyard manure and soil [23].

8.2.2 IDENTIFICATION OF TRB IN THE SLUDGE

According to the National Antimicrobial Resistance Monitoring System, bacteria are identified as TRB if they can survive under 20 mg/L tetracycline stress [8,10,12]. In this study, the bacteria with successive increases of relative abundance in response of tetracycline stress enhancement were considered TRB.

In the sludge cultured with 20 mg/L tetracycline, TRB consisted of *Proteobacteria, Armatimonadetes, Verrucomicrobia* and *Chloroflexi* phyla, accounting for 60.61%, 16.97%, 15.76% and 6.67% of the total TRB community, respectively (Table 2). At the level of class, TRB consisted of *BetaProteobacteria, AlphaProteobacteria, Armatimonadia, Verrucomicrobiae* and *Anaerolineae*, among which *BetaProteobacteria* and *AlphaProteobacteria* were the main classes. This is supported by a previous study indicating that *AlphaProteobacteria* and *BetaProteobacteria* dominated in an oxytetracycline production wastewater treatment plant [24] and an aerobic reactor treating high-concentration antibiotic wastewater [25]. A total of nine genera were identified for TRB in the sludge, among which *Sulfuritalea* (0.54%) had the highest abundance, followed by *Armatimonas* (0.39%), *Prosthecobacter* (0.37%), *Hyphomicrobium* (0.34%), *Azonexus* (0.20%), *Longilinea* (0.15%), *Novosphingobium* (0.13%), *Paracoccus* (0.11%) and *Rhodobacter* (0.10%) (Table 2).

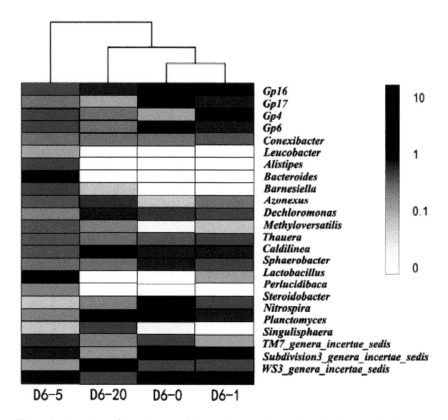

Figure 2. Heat map of genera occurring at >1% abundance in at least one sludge sample. Scale bar on the right shows the variation of the normalized abundance of the genera. D6-0, D6-1, D6-5 and D6-20: sludge cultured with 0, 1, 5 and 20 mg/L tetracycline for 6 days, respectively.

Paracoccus serving as an important denitrifier [26] has been reported to be TRB in STP sludge treated with tetracycline [13]. To our knowledge, the other 8 genera were firstly identified as TRB in this study, indicating that pyrosequencing is a new powerful tool to profile the ARB communities in the environment. An in-depth investigation showed that the function of the newly identified TRB mainly included denitrification and degradation. The TRB genus *Sulfuritalea* dominating in the sludge is a

TABLE 2. Taxon composition profile of tetracycline resistant bacteria (TRB) in activated sludge.

Phylum	Class	Genus	D6-0	D6-1	D6-5	D6-20
Proteobacteria	Betaproteobac-teria	Sulfuritalea	0.01%	0.10%	0.10%	0.54%
Armatimonadetes	Armatimonadia	Armatimonas	ND	ND	0.01%	0.39%
Verrucomicrobia	Verrucomicro-biae	Prosthecobacter	ND	ND	0.01%	0.37%
Proteobacteria	Alphaproteo-bacteria	Hyphomicro-bium	ND	0.11%	0.13%	0.34%
Proteobacteria	Betaproteobac-teria	Azonexus	0.01%	0.15%	0.18%	0.20%
Chloroflexi	Anaerolineae	Longilinea	ND	0.01%	0.01%	0.15%
Proteobacteria	Alphaproteo-bacteria	Novosphin-gobium	0.04%	0.04%	0.07%	0.13%
Proteobacteria	Alphaproteo-bacteria	Paracoccus	ND	0.01%	0.01%	0.11%
Proteobacteria	Alphaproteo-bacteria	Rhodobacter	ND	0.01%	0.03%	0.10%

ND: Not detectable; D6-0, D6-1, D6-5 and D6-20: sludge cultured with 0, 1, 5 and 20 mg/L tetracycline for 6 days, respectively.

denitrifier frequently detected in freshwater lakes [27]. Both nitrate-reduction bacterium *Azonexus caeni* [28] and denitrifying photosynthetic bacteria *Rhodobacter* [29] were previously isolated from sludge of wastewater treatment plants. *Hyphomicrobium* sp. can grow on media with chloromethane, methanol, methylamine and ethanol as sole carbon and energy sources, and the microorganism has been used for bioremediation of gasoline-contaminated site [30]. *Novosphingobium* sp. is widely distributed in the environment, e.g., groundwater treatment bioreactor [31], deep-sea environment [32] and freshwater lakes [33], and can degrade various aromatic compounds including polychlorophenol and polycyclic aromatic hydrocarbons. In addition, it has been reported that *Novosphingobium* sp.

isolated from activated sludge of a Japanese STP is capable of estradiol degradation [34].

8.2.3 EFFECTS OF TETRACYCLINE STRESS
ON THE ABUNDANCE AND DIVERSITY OF ARGS

To investigate the impact of tetracycline stress on the abundance of ARGs, the sludge samples fed with 0 and 20 mg/L tetracycline were selected for Illumina high-throughput sequencing. Alignments of the Illumina reads against the Antibiotic Resistance Genes Database (ARDB) showed that a total of 2168 reads (0.0192%) from the sludge under no tetracycline stress and 515 reads (0.0043%) from the sludge fed with 20 mg/L tetracycline were annotated as 47 and 41 types of known ARGs, respectively (Figure 3). As a common hypothesis, ARGs may have higher abundance in the presence of antibiotics [35].

However, this study showed that tetracycline treatment decreased both the occurrence and diversity of non-tetracycline ARGs, although the abundance of tet genes increased. The considerable decrease in the abundance of sulfonamide resistance gene sul2 (from 83.49% to 14.76%) mainly contributed to the diminishment of the non-tetracycline ARGs (Figure 3). Table S1 shows that sul2 (ARDB accession number: CAE53425) dominated in the sludge without tetracycline stress, but had much lower abundance in the sludge fed with 20 mg/L tetracycline (Table S2). *Pasteurella multocida* was often found to carry sul2 (http://ardb.cbcb.umd.edu/cgi/search.cgi?db=L&field=ni&term=CAE53425), and its growth can be inhibited by tetracycline via interference with protein synthesis by binding to the bacterial 30S ribosomal subunit [36]. Interestingly, this study showed that *Salmonella enterica* plasmid pCVM19633_110 and *Pasteurella multocida* plasmid pCCK381 predominant in the sludge under no tetracycline stress (Table S3) had much lower abundance in the sludge fed with 20 mg/L tetracycline (Table S4). This may be also responsible for the reduction of sul2 abundance induced by tetracycline treatment since sul2 are located on the genomes of the two plasmids [35]. In addition, we summarized the types of the ARGs detected in the sludge separately fed with 0 and 20 mg/L tetracycline (Figure 4A). Most of the assigned sequencing reads

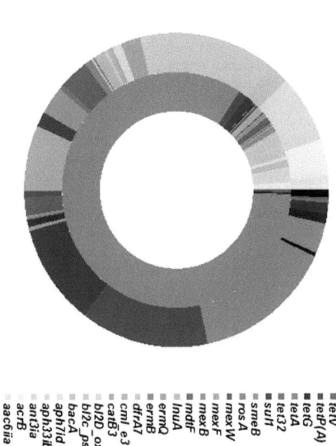
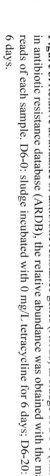

Figure 3. Relative abundance of antibiotic resistance genes (ARGs) in sludge of D6-0 (Inner ring) and D6-20 (Outer ring). After searching in antibiotic resistance database (ARDB), the relative abundance was obtained with the matched sequencing reads normalized to the total reads of each sample. D6-0: sludge incubated with 0 mg/L tetracycline for 6 days; D6-20: sludge incubated with 20 mg/L tetracycline for 6 days.

were found to be involved in sulfonamide resistance in the sludge containing no tetracycline (84.78%) and the sludge fed with 20 mg/L tetracycline
(28.35%). Figure 4A illustrates that the tetracycline selection pressure (20
mg/L) promoted multidrug, aminoglycoside and tetracycline (from 0.78%
to 6.99%) resistances in the sludge. A previous study also indicated that
incubation in the presence of tetracycline favored the emergence of multi-
drug-resistance mutants in *Pseudomonas aeruginosa* [37].

PCRs showed that 11 tet genes among the 15 tested genes were present
in the sludge, including tetA, tetB, tetC, tetG, tetK and tetP(A) encoding
tetracycline efflux proteins, tetM, tetO, tetS and tetW encoding ribosomal
protection proteins, and tetX encoding enzymatic modification protein
(Figure S1). Previous studies showed that tetA, tetC and tetG were more
abundant than other tet genes [11,38], so tetA, tetC and tetG were selected
for qPCR to investigate the impact of tetracycline stress on the abundance
of tet genes. TetC had higher abundance than tetA and tetG by an order of
magnitude, which indicated that tetC might play an important role in tetracycline resistance in the STP sludge. After 6 days treatment, the sludge
fed with 5 or 20 mg/L tetracycline had comparatively higher levels of tetA,
tetC and tetG, but 1 mg/L tetracycline treatment posed no evident effect
($p > 0.05$) (Figure 5). The result is confirmed by metagenomic analysis
showing that tetA, tetC and tetG genes increased from 0.09% to 3.11%,
0.18% to 0.97% and 0.09% to 1.17% after 6 days treatment with 20 mg/L
tetracycline, respectively (Figure 4B). Enhancement of tetracycline concentration led to significant increase in the relative abundance of tetA, tetC
and tetG (Figure 4B and Figure 5), which may result from the microbial
community shift (Figure 2). Li et al. [13] reported a similar result that tetA
and tetG significantly increased after tetracycline treatment. In this study,
metagenomic analysis also indicated that the relative abundance of tetO,
tetP(A), tetW, tetX and tet33 increased after tetracycline incubation, but
some minor tet genes, e.g., tetP(B), tetV and tet32, had lower abundance in
response of tetracycline treatment (Figure 4B). Zhang et al. [14] indicated
that proliferation of the ARGs can be accelerated in the activated sludge
under tetracycline pressure. qPCR results showed that the tet genes tended
to have the highest abundance under the condition of 5 mg/L tetracycline
(Figure 5), which is confirmed by the pyrosequencing results demonstrating that the bacterial community structure of the sludge treated with 5

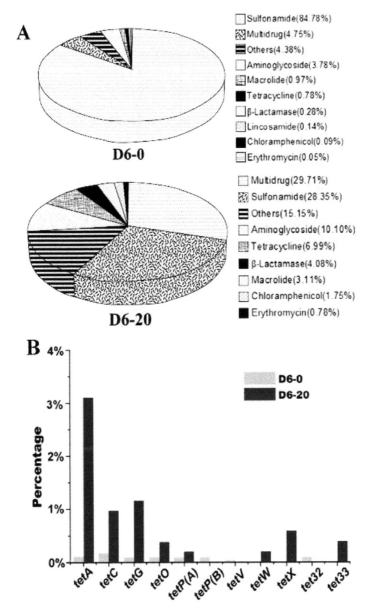

Figure 4. Antibiotic resistance patterns (A) and tet genes (B) in sludge treated with 0 mg/L (D6-0) and 20 mg/L (D6-20) tetracycline. The resistance genes identified were grouped according to antibiotic types after alignment of the high-throughput sequencing reads against antibiotic resistance database (ARDB). D6-0: sludge incubated with 0 mg/L tetracycline for 6 days; D6-20: sludge incubated with 20 mg/L tetracycline for 6 days.

mg/L tetracycline were evidently divergent from those of the other three sludge samples (Figure 2). Further, it is known that antibiotic treatment at subinhibitory concentrations can increase the rate of mutation, horizontal gene transfer and spread of antibiotic resistance [39].

Most of the functional genes are considered conserved, but a previous study [11] showed that tetG had an extremely high diversity in STPs. In this study, a total of 52 clones, including 26 clones from the sludge containing no tetracycline and 26 clones from the sludge incubated with 20 mg/L tetracycline, were selected to investigate the effect of tetracycline on diversity of tetG. Results showed that 19 tetG genotypes occurred in the sludge under no tetracycline stress and 21 genotypes were present in the sludge treated with 20 mg/L tetracycline.

Among the clones of sludge fed with no tetracycline, types G0-1 and G0-26 had a 100% identity to *Salmonella typhimurium* tetG (Y19117.1), and G0-9 were identified as the corresponding sequence of *Stenotrophomonas* sp. tetG (EF055281.1). Each of the types G0-8 and G0-25 had a similarity of 95% to the most closely related known gene: *Mannheimia haemolytica* tetG (AJ276217.1), while G0-5 and G0-22 had a sequence identity of only 94% to *Ochrobactrum* sp. tetG (EF055280.1) (Figure 6A). Tetracycline may increase the diversity of tetG, since five genotypes of tetG cloned from the sludge treated with 20 mg/L tetracycline could not be matched to the known tetG genes deposited in GenBank (Figure 6B). The selective pressure resulted from absorption of tetracycline by activated sludge may contribute to alterations on tetG DNA sequences [40].

8.3 MATERIALS AND METHODS

8.3.1 BATCH EXPERIMENTS

Untreated sewage wastewater (25 L) and activated sludge (5 L) were sampled from Jiangxinzhou STP (Nanjing, China). Water and sludge samples were transported on ice to lab within 2 h. Sewage was filtered by 0.45 μm nitrate cellulose membrane. Activated sludge was centrifuged at 4000 rpm for 10 min under 4 °C, and the pellets were dissolved in 1 L of the filtered sewage water. In the batch assay, four 500 mL glass flasks with activated

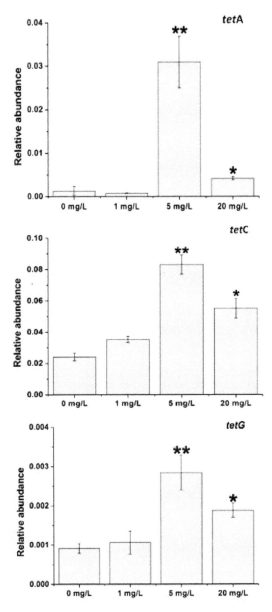

Figure 5. Relative abundance of tetA, tetC and tetG in sludge fed with different concentrations of tetracycline for 6 days. qPCR was used to determine the relative abundance normalized to the total copy number of 16S rRNA genes in corresponding samples. ** $p < 0.01$, comparing 5 mg/L with 0 mg/L; * $p < 0.05$, comparing 20 mg/L with 0 mg/L.

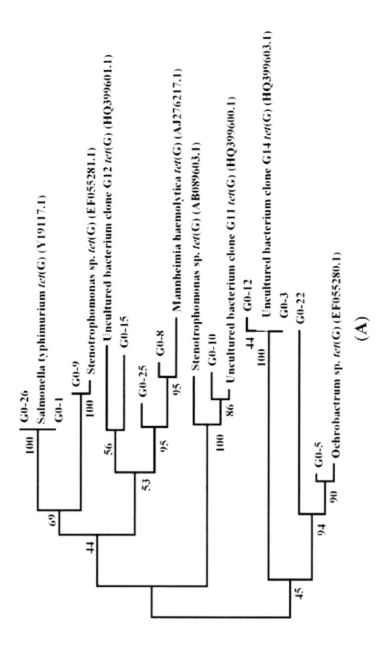

Figure 6. Relative abundance of tetA, tetC and tetG in sludge fed with different concentrations of tetracycline for 6 days. qPCR was used to determine the relative abundance normalized to the total copy number of 16S rRNA genes in corresponding samples. ** p < 0.01, comparing 5 mg/L with 0 mg/L; * p < 0.05, comparing 20 mg/L with 0 mg/L.

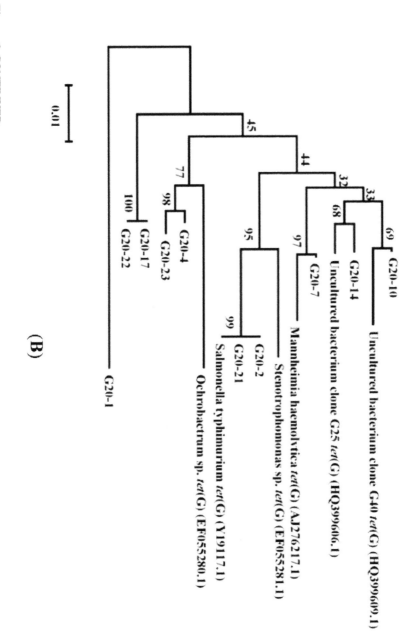

Figure 6. CONTINUED.

sludge mixed liquor (300 mL) separately containing 0, 1, 5 and 20 mg/L tetracycline [8,10,12,13] were run continuously at (26 ± 1) °C for 6 days. The glass flasks were cultured in a shaking incubator at 140 rpm for the aeration and mixing, and the incubator was covered with aluminum foil to avoid possible tetracycline photolysis. Every 24 h for 6 days, one third (100 mL, v/v) of the sludge samples were taken out from the reactors for biomass determination and DNA extraction. The remaining mixed liquor (200 mL) was centrifuged at 4000 rpm for 10 min under 4 °C. The pellets were transferred back to the reactors and then re-suspended with 300 mL filtered sewage containing corresponding concentrations of tetracycline (0, 1, 5 or 20 mg/L) for subsequent 24-h culture.

8.3.2 DNA EXTRACTION AND PCR

For DNA protection, activated sludge was sampled from the reactors every 24 h and mixed with 100% ethanol immediately at a ratio of 1:1 (v/v). The mixture was centrifuged at 4000 rpm for 10 min under 4 °C to collect the pellets (approximately 200 mg) for DNA extraction. The total DNA extraction was conducted using the FastDNA SPIN Kit for Soil (MP Biomedicals, Santa Ana, CA, USA). The DNA concentrations and purity were determined through microspectrophotometry (NanoDrop®ND-2000, NanoDrop Technologies, Willmington, DE, USA). The DNA products were stored at −20 °C until further molecular analyses.

According to the previous studies, 15 tet genes were amplified using primers listed in Table S5 in a 30-μL reaction mixture containing 1× PCR buffer, 100 μM dNTP, 2 pmol of each primer, 150 ng of template DNA and 1 U of EX Taq polymerase (TaKaRa, Shiga, Japan). TetO and tetS were amplified using the following conditions: initial denaturation at 95 °C for 7 min, followed by 40 cycles of 94 °C for 15 s, 50.3 °C (tetO) or 56 °C (tetS) for 30 s and 72 °C for 30 s, with a final extension of 72 °C for 7 min. PCR of tetW was carried out through: initial denaturation at 94 °C for 5 min, followed by 25 cycles of 94 °C for 30 s, 64 °C for 30 s and 72 °C for 30 s, with a final extension of 72 °C for 7 min. For the other 12 tet genes, PCR amplification was conducted according to the following protocols: initial denaturation at 94 °C for 5 min, followed by 35 cycles of denaturation at 94 °C for 1 min, annealing for 1 min at different temperatures (Table S5) and extension at 72 °C for 1.5 min, with a final elongation step at 72 °C for 10 min. The PCR products obtained were analyzed by gel electrophoresis using 1% (w/v) agarose in 1× TAE buffer and further confirmed by DNA sequencing. To check reproducibility, duplicate PCR reactions were performed for each sample and sterile water was used as the negative control. The PCR products (longer than 200 bp) sequencing data were deposited in NCBI under accession number KJ603161~KJ603167.

8.3.3 QUANTITATIVE REAL-TIME PCR

TetA, tetC and tetG were selected for quantitative assay using SYBR Green I qPCR. The plasmids containing target genes were obtained by molecular cloning. In detail, the PCR products of each tet gene were purified using the DNA Fragment Purification Kit (TaKaRa, Shiga, Japan) and cloned into the pMD19-T Vector (TaKaRa, Shiga, Japan). Plasmids carrying each tet gene were extracted and purified using the MiniBest Plasmid Purification Kit (TaKaRa, Shiga, Japan). Plasmid concentrations were determined by NanoDrop®ND-2000 (NanoDrop Technologies, Willmington, DE, USA). qPCRs were performed in 96-well plates with a final volume of 20 μL containing SYBR Premix EX Taq (TaKaRa, Shiga, Japan) super mix (10 μL), 10 μM primer (0.2 μL each), DNA templates (8 μL) and ddH$_2$O (1.6 μL). Thermal cycling and fluorescence detection were conducted in Corbett Real-Time PCR with the Rotor-Gene 6000 Series Software 1.7 (QIAGEN, Nijmegen Area, The Netherlands). qPCR were performed using the following protocol: 94 °C for 3 min, followed by 40 cycles of 94 °C for 30 s, annealing at different temperatures (Table S5) for 45 s, and extension at 72 °C for 45 s. Each reaction was run in triplicate.

Five to seven-point calibration curves (C$_t$ value versus log of initial tet gene copy) were generated for qPCR using 10-fold serial dilution of the tet-carrying plasmids. The PCR efficiency of each gene ranged from

92.1% to 102.6% with R^2 values more than 0.995 for all calibration curves. Based on the calibration curves, the abundance of tet genes were calculated through the Ct values of the experimental samples. To minimize the potential variations in extraction efficiencies, eubacterial 16S rRNA genes were quantified using the method recommended by López-Gutiérrez et al. [41], and the relative abundance of tet genes was normalized to the total bacterial community.

8.3.4 CLONING AND PHYLOGENETIC ANALYSIS OF TETG

The PCR products of tetG were cloned to investigate the diversity of the genes in the sludge treated with 0 and 20 mg/L for 6 days. The purified PCR products were cloned to pMD19-T Vector (TaKaRa, Shiga, Japan). A total of 52 clones were randomly selected for the library construction, sequencing and subsequent similarity analysis. Nucleotide sequences of tetG were aligned using CLUSTALW [42]. The clones sharing a consensus sequence were grouped into one genotype, and only one representative in each group was selected for construction of phylogenetic trees. The GenBank sequences having the highest identity to the sequences obtained in this study were retrieved for phylogenetic trees construction. The neighbor-joining trees were constructed using Molecular Evolutionary Genetics Analysis (MEGA version 5.05) [42] and bootstrap analysis with 1000 replicates was conducted to evaluate the significance of the nodes. The 52 sequences of tetG cloning obtained in this study have been deposited in NCBI (Accession number: KJ603168~KJ603219).

8.3.5 454 PYROSEQUENCING

The DNA extracted from activated sludge dosed with different levels of tetracycline (0, 1, 5 and 20 mg/L) for 6 days were subjected to Beijing Genome Institute (Shenzhen, China) for 16S rRNA gene pyrosequencing. The primers, V3F (5'-ACTCCTACGGGAGGCAGCAG-3') and

V4R (5'-TACNVGGGTATCTAATCC-3'), targeting the hypervariable V3–V4 region (about 460 bp) were used to amplify the bacterial 16S rRNA gene. PCRs were conducted in a reaction system (50 μL) containing 1× Pfx Amplification Buffer (Invitrogen, Carlsbad, CA, USA), 0.4 mM dNTP, 2 mM MgSO4, 0.4 μM each fusion primer, 1 μL of template DNA and 2 U of Platinum® Pfx DNA Polymerase (Invitrogen, Carlsbad, CA, USA). The 10 nucleotide "barcode" was permuted for each sample to separate the corresponding reads from the data pool generated in a single pyrosequencing run. PCR amplification was performed according to the following protocols: initial denaturation at 94 °C for 3 min, followed by 30 cycles of 94 °C for 30 s, 62 °C for 30 s and 70 °C for 45 s, with a final elongation step at 70 °C for 7 min. In order to minimize the impact of potential early-round PCR errors, amplicon libraries were prepared by a cocktail of three independent PCR products for each sample. The PCR products were purified using QIAquick PCR Purification Kit (Qiagen, Hilden, Germany) and quantified with an Agilent 2100 Bioanalyzer (Agilnet, Santa Clara, CA, USA). Equal DNA mass of each purified amplicon library from different samples were mixed for pyrosequencing on the Roche 454 FLX Titanium platform (Roche, Indianapolis, IN, USA) at Beijing Genome Institute (Shenzhen, China). The sequencing data were deposited in NCBI Sequence Read Archive under accession number SRP035342.

8.3.6 ILLUMINA HIGH-THROUGHPUT SEQUENCING

The metagenomic DNA extracted from the sludge cultured with 0 and 20 mg/L tetracycline was individually subjected to high-throughput sequencing using Illumina Hiseq 2000 (Illumina, San Diego, CA, USA) according to the manufacturer's instructions. The "Index 101 PE" (Paired End sequencing, 101-bp reads and 8-bp index sequence) sequencing strategy was used for the high-throughput sequencing, which generates nearly equal amount of clean reads for each sample. A base-calling pipeline (Sequencing Control Software, Illumina, San Diego, CA, USA) was applied to process the raw fluorescent images and the call sequences. The raw reads containing three or more "N" or contaminated by adapter (>15 bp overlap) were removed, and the filtered clean reads (about 1.6 Gb per each sample) were used for further metagenomic analyses. The sequencing data were deposited in the metagenomics RAST server (MG-RAST) [43] under accession number 4494851.3 (sludge treated with 20 mg/L tetracycline) and 4494856.3 (sludge without tetracycline treatment).

8.3.7 BIOINFORMATICS ANALYSIS

After 454 pyrosequencing, all the reads were subjected to the Pyrosequencing Pipeline Initial Process [44] of the Ribosomal Database Project (RDP): (1) To sort the reads to the designated sample based on their nucleotide barcode; (2) To trim off the adapters and barcodes using the default parameters; and (3) To remove sequences containing ambiguous "N" or shorter than 200 bp [45]. Sequences were de-noised using the "pre.cluster" command in the Mothur platform to remove the erroneous sequences due to pyrosequencing errors [46,47]. PCR chimeras were filtered out using Chimera Slayer [48]. The reads marked as chimeras were extracted and submitted to RDP. Those being assigned to any known genus with 90% confidence were integrated with the non-chimera reads [49], to form the "effective sequences". The effective sequences of each sample were resubmitted to the RDP Classifier [50] to identify the archaeal and bacterial sequences, and the unexpected archaeal sequences were manually removed. In order to study the tetracycline effect on microbial communities, the samples of day 6 were individually selected for pyrosequencing, which generated a total of 42,556 reads for the four samples. As shown in Table 1, low quality reads were filtered using RDP and the effective reads were obtained after trimming the adapters, barcodes and primers. After denoising, filtering out chimeras and removing the archaeal sequences, the minimum number of bacterial sequences in the four samples was 7097. To fairly compare the four samples at the same sequencing depth, the number of the sequences from each sample was normalized to be 7097 for subsequent bioinformatics analyses. Taxonomic assignment of the sequences was separately performed using the RDP's Classifier. A bootstrap cutoff of 80% was applied to assign the sequences to different taxonomy levels. Richness and diversity indices including OTUs, Chao 1 estimator and Shannon index, as well as rarefaction curves, were calculated using the relevant RDP modules, including Rarefaction and Shannon & Chao1 index.

Illumina sequencing reads were aligned against a self-established database via off-line BLAST to identify ARGs and plasmids in the sludge samples. A protein database of ARGs were created by downloading all sequences in ARDB (7828 sequences) [51]. A read was identified as an ARG according to its best BLAST hit (blastx) if the similarity was above 90% and the alignments was at least 25 amino acids [35]. The nucleotide sequences of plasmids were downloaded from NCBI RefSeq database (2408 plasmid genome sequences). A read was annotated as plasmids if the best BLAST hits (blastn) had a nucleotide sequence similarity >95% over at least 90 bp alignment [35].

8.4 CONCLUSIONS

Tetracycline treatment can affect bacterial community structure and increase total abundance and diversity of tet genes in the STP sludge, but tends to reduce the abundance of sul2 predominant in the sludge without tetracycline treatment. Several genera of TRB, including *Sulfuritalea, Armatimonas, Prosthecobacter, Hyphomicrobium, Azonexus, Longilinea, Paracoccus, Novosphingobium* and *Rhodobacter* are present in the sludge. Comparatively, antibiotic treatment at subinhibitory concentrations can pose greater effects on the bacterial community composition. The microbial community shift may be responsible for the ARGs distribution patterns variation induced by the tetracycline treatment. As a culture-independent method, pyrosequencing of 16S rRNA gene provides a comprehensive insight into microbial community structure of ARB. Illumina high-throughput sequencing offers enough sequencing depth for metagenomic analysis of ARGs. Combined use of 454 pyrosequencing and Illumina high-throughput sequencing is considered a promising tool for exploration of ARB and ARGs in the environment.

REFERENCES

1. Kümmerer, K. Resistance in the environment. J. Antimicrob. Chemoth. 2004, 54, 311–320.
2. Levy, S.B.; Marshall, B. Antibacterial resistance worldwide: Causes, challenges and responses. Nat. Med. 2004, 10, 122–129.
3. Walsh, T.R.; Weeks, J.; Livermore, D.M.; Toleman, M.A. Dissemination of NDM-1 positive bacteria in the New Delhi environment and its implications for human health: An environmental point prevalence study. Lancet Infect. Dis. 2011, 11, 355–362.
4. Sarmah, A.K.; Meyer, M.T.; Boxall, A. A global perspective on the use, sales, exposure pathways, occurrence, fate and effects of veterinary antibiotics (VAs) in the environmt. Chemosphere 2006, 65, 725–759.
5. Kim, H.; Hong, Y.; Park, J.-E.; Sharma, V.K.; Cho, S.-I. Sulfonamides and tetracyclines in livestock wastewater. Chemosphere 2013, 91, 888–894.
6. Ben, W.; Qiang, Z.; Adams, C.; Zhang, H.; Chen, L. Simultaneous determination of sulfonamides, tetracyclines and tiamulin in swine wastewater by solid-phase extraction and liquid chromatography-mass spectrometry. J. Chromatogr. A 2008, 1202, 173–180.
7. Li, B.; Zhang, T.; Xu, Z.; Fang, H.H.P. Rapid analysis of 21 antibiotics of multiple classes in municipal wastewater using ultra performance liquid chromatography-tandem mass spectrometry. Anal. Chim. Acta 2009, 645, 64–72.
8. Gao, P.; Munir, M.; Xagoraraki, I. Correlation of tetracycline and sulfonamide antibiotics with corresponding resistance genes and resistant bacteria in a conventional municipal wastewater treatment plant. Sci. Total Environ. 2012, 421, 173–183.
9. Taviani, E.; Ceccarelli, D.; Lazaro, N.; Bani, S.; Cappuccinelli, P.; Colwell, R.R.; Colombo, M.M. Environmental Vibrio spp., isolated in Mozambique, contain a polymorphic group of integrative conjugative elements and class 1 integrons. FEMS Microbiol. Ecol. 2008, 64, 45–54.
10. Zhang, T.; Zhang, M.; Zhang, X.-X.; Fang, H.H. Tetracycline resistance genes and tetracycline resistant lactose-fermenting Enterobacteriaceae in activated sludge of sewage treatment plants. Environ. Sci. Technol. 2009, 43, 3455–3460.
11. Zhang, X.-X.; Zhang, T. Occurrence, abundance, and diversity of tetracycline resistance genes in 15 sewage treatment plants across China and other global locations. Environ. Sci. Technol. 2011, 45, 2598–2604.
12. Kim, S.; Jensen, J.N.; Aga, D.S.; Weber, A.S. Tetracycline as a selector for resistant bacteria in activated sludge. Chemosphere 2007, 66, 1643–1651.
13. Li, B.; Zhang, X.-X.; Guo, F.; Wu, W.; Zhang, T. Characterization of tetracycline resistant bacterial community in saline activated sludge using batch stress incubation with high-throughput sequencing analysis. Water Res. 2013, 47, 4207–4216.
14. Zhang, W.; Huang, M.-H.; Qi, F.-F.; Sun, P.-Z.; van Ginkel, S.W. Effect of trace tetracycline concentrations on the structure of a microbial community and the development of tetracycline resistance genes in sequencing batch reactors. Bioresour. Technol. 2013, 150, 9–14.

15. Walsh, F.; Cooke, N.M.; Smith, S.G.; Moran, G.P.; Cooke, F.J.; Ivens, A.; Wain, J.; Rogers, T.R. Comparison of two DNA microarrays for detection of plasmid-mediated antimicrobial resistance and virulence factor genes in clinical isolates of Enterobacteriaceae and non-Enterobacteriaceae. Int. J. Antimicrob. Agent 2010, 35, 593–598.

16. Roesch, L.F.; Fulthorpe, R.R.; Riva, A.; Casella, G.; Hadwin, A.K.; Kent, A.D.; Daroub, S.H.; Camargo, F.A.; Farmerie, W.G.; Triplett, E.W. Pyrosequencing enumerates and contrasts soil microbial diversity. ISME J. 2007, 1, 283–290.

17. Qin, J.; Li, R.; Raes, J.; Arumugam, M.; Burgdorf, K.S.; Manichanh, C.; Nielsen, T.; Pons, N.; Levenez, F.; Yamada, T. A human gut microbial gene catalogue established by metagenomic sequencing. Nature 2010, 464, 59–65.

18. Håvelsrud, O.; Haverkamp, T.; Kristensen, T.; Jakobsen, K.; Rike, A. Metagenomic and geochemical characterization of pockmarked sediments overlaying the Troll petroleum reservoir in the North Sea. BMC Microbiol. 2012, 12, 203.

19. Ye, L.; Zhang, T.; Wang, T.; Fang, Z. Microbial structures, functions, and metabolic pathways in wastewater treatment bioreactors revealed using high-throughput sequencing. Environ. Sci. Technol. 2012, 46, 13244–13252.

20. Li, D.; Qi, R.; Yang, M.; Zhang, Y.; Yu, T. Bacterial community characteristics under long-term antibiotic selection pressures. Water Res. 2011, 45, 6063–6073.

21. Czárán, T.L.; Hoekstra, R.F.; Pagie, L. Chemical warfare between microbes promotes biodiversity. Proc. Natl. Acad. Sci. USA 2002, 99, 786–790.

22. Lozada, M.; Figuerola, E.L.M.; Itria, R.F.; Erijman, L. Replicability of dominant bacterial populations after long-term surfactant-enrichment in lab-scale activated sludge. Environ. Microbiol. 2006, 8, 625–638.

23. Kobashi, Y.; Hasebe, A.; Nishio, M.; Uchiyama, H. Diversity of tetracycline resistance genes in bacteria isolated from various agricultural environments. Microbes Environ. 2007, 22, 44–51.

24. Li, D.; Yu, T.; Zhang, Y.; Yang, M.; Li, Z.; Liu, M.; Qi, R. Antibiotic resistance characteristics of environmental bacteria from an oxytetracycline production wastewater treatment plant and the receiving river. Appl. Environ. Microb. 2010, 76, 3444–3451.

25. Deng, Y.; Zhang, Y.; Gao, Y.; Li, D.; Liu, R.; Liu, M.; Zhang, H.; Hu, B.; Yu, T.; Yang, M. Microbial community compositional analysis for series reactors treating high level antibiotic wastewater. Environ. Sci. Technol. 2011, 46, 795–801.

26. Vacková, L.; Srb, M.; Stloukal, R.; Wanner, J. Comparison of denitrification at low temperature using encapsulated Paracoccus denitrificans, Pseudomonas fluorescens and mixed culture. Bioresour. Technol. 2011, 102, 4661–4666.

27. Kojima, H.; Fukui, M. Sulfuritalea hydrogenivorans gen. nov., sp. nov., a facultative autotroph isolated from a freshwater lake. Int. J. Syst. Evol. Microbiol. 2011, 61, 1651–1655.

28. Quan, Z.-X.; Im, W.-T.; Lee, S.-T. Azonexus caeni sp. nov., a denitrifying bacterium isolated from sludge of a wastewater treatment plant. Int. J. Syst. Evol. Microbiol. 2006, 56, 1043–1046.

29. Kim, J.K.; Lee, B.-K.; Kim, S.-H.; Moon, J.-H. Characterization of denitrifying photosynthetic bacteria isolated from photosynthetic sludge. Aquacult. Eng. 1999, 19, 179–193.

30. McDonald, I.R.; Doronina, N.V.; Trotsenko, Y.A.; McAnulla, C.; Murrell, J.C. Hyphomicrobium chloromethanicum sp. nov. and Methylobacterium chloromethanicum sp. nov., chloromethane-utilizing bacteria isolated from a polluted environment. Int. J. Syst. Evol. Microbiol. 2001, 51, 119–122.

31. Tiirola, M.A.; Männistö, M.K.; Puhakka, J.A.; Kulomaa, M.S. Isolation and characterization of Novosphingobium sp. strain MT1, a dominant polychlorophenol-degrading strain in a groundwater bioremediation system. Appl. Environ. Microbiol. 2002, 68, 173–180.

32. Yuan, J.; Lai, Q.; Zheng, T.; Shao, Z. Novosphingobium indicum sp. nov., a polycyclic aromatic hydrocarbon-degrading bacterium isolated from a deep-sea environment. Int. J. Syst. Evol. Microbiol. 2009, 59, 2084–2088.

33. Liu, Z.-P.; Wang, B.-J.; Liu, Y.-H.; Liu, S.-J. Novosphingobium taihuense sp. nov., a novel aromatic-compound-degrading bacterium isolated from Taihu Lake, China. Int. J. Syst. Evol. Microbiol. 2005, 55, 1229–1232.

34. Fujii, K.; Satomi, M.; Morita, N.; Motomura, T.; Tanaka, T.; Kikuchi, S. Novosphingobium tardaugens sp. nov., an oestradiol-degrading bacterium isolated from activated sludge of a sewage treatment plant in Tokyo. Int. J. Syst. Evol. Microbiol. 2003, 53, 47–52.

35. Kristiansson, E.; Fick, J.; Janzon, A.; Grabic, R.; Rutgersson, C.; Weijdegård, B.; Söderström, H.; Larsson, D.J. Pyrosequencing of antibiotic-contaminated river sediments reveals high levels of resistance and gene transfer elements. PLoS One 2011, 6, 1–7.

36. Stevens, D.L.; Higbee, J.W.; Oberhofer, T.R.; Everett, E.D. Antibiotic susceptibilities of human isolates of Pasteurella. multocida. Antimicrob. Agents Chemother. 1979, 16, 322–324.

37. Alonso, A.; Campanario, E.; Martinez, J.L. Emergence of multidrug-resistant mutants is increased under antibiotic selective pressure in Pseudomonas aeruginosa. Microbiol-Sgm 1999, 145, 2857–2862.

38. Auerbach, E.A.; Seyfried, E.E.; McMahon, K.D. Tetracycline resistance genes in activated sludge wastewater treatment plants. Water Res. 2007, 41, 1143–1151.

39. Laureti, L.; Matic, I.; Gutierrez, A. Bacterial responses and genome instability induced by subinhibitory concentrations of antibiotics. Antibiotics 2013, 2, 100–114.

40. Li, B.; Zhang, T. Biodegradation and adsorption of antibiotics in the activated sludge process. Environ. Sci. Technol. 2010, 44, 3468–3473.

41. López-Gutiérrez, J.C.; Henry, S.; Hallet, S.; Martin-Laurent, F.; Catroux, G.; Philippot, L. Quantification of a novel group of nitrate-reducing bacteria in the environment by real-time PCR. J. Microbiol. Meth. 2004, 57, 399–407.

42. Tamura, K.; Peterson, D.; Peterson, N.; Stecher, G.; Nei, M.; Kumar, S. MEGA5: molecular evolutionary genetics analysis using maximum likelihood, evolutionary distance, and maximum parsimony methods. Mol. Biol. Evol. 2011, 28, 2731–2739.

43. Meyer, F.; Paarmann, D.; D'Souza, M.; Olson, R.; Glass, E.M.; Kubal, M.; Paczian, T.; Rodriguez, A.; Stevens, R.; Wilke, A.; et al. The metagenomics RAST server–a public resource for the automatic phylogenetic and functional analysis of metagenomes. BMC Bioinform. 2008, 9, 1–8.

44. Cole, J.R.; Wang, Q.; Cardenas, E.; Fish, J.; Chai, B.; Farris, R.J.; Kulam-Syed-Mohideen, A.S.; McGarrell, D.M.; Marsh, T.; Garrity, G.M. The ribosomal database

project: Improved alignments and new tools for rRNA analysis. Nucleic Acids Res. 2009, 37, 141–145.

45. Claesson, M.J.; O'Sullivan, O.; Wang, Q.; Nikkila, J.; Marchesi, J.R.; Smidt, H.; de Vos, W.M.; Ross, R.P.; O'Toole, P.W. Comparative analysis of pyrosequencing and a phylogenetic microarray for exploring microbial community structures in the human distal intestine. PLoS One 2009, 4, 1–15.

46. Huse, S.M.; Welch, D.M.; Morrison, H.G.; Sogin, M.L. Ironing out the wrinkles in the rare biosphere through improved OTU clustering. Environ. Microbiol. 2010, 12, 1889–1898.

47. Roeselers, G.; Mittge, E.K.; Stephens, W.Z.; Parichy, D.M.; Cavanaugh, C.M.; Guil-lemin, K.; Rawls, J.F. Evidence for a core gut microbiota in the zebrafish. ISME J. 2011, 5, 1595–1608.

48. Haas, B.J.; Gevers, D.; Earl, A.M.; Feldgarden, M.; Ward, D.V.; Giannoukos, G.; Ci-ulla, D.; Tabbaa, D.; Highlander, S.K.; Sodergren, E. Chimeric 16S rRNA sequence formation and detection in Sanger and 454-pyrosequenced PCR amplicons. Genome Res. 2011, 21, 494–504.

49. Zhang, T.; Shao, M-F.; Ye, L. 454 Pyrosequencing reveals bacterial diversity of acti-vated sludge from 14 sewage treatment plants. ISME J. 2012, 6, 1137–1147.

50. Wang, Q.; Garrity, G.M.; Tiedje, J.M.; Cole, J.R. Naive bayesian classifier for rapid assignment of rRNA sequences into the new bacterial taxonomy. Appl. Environ. Microbiol. 2007, 73, 5261–5267.

51. Liu, B.; Pop, M. ARDB-Antibiotic Resistance Genes Database. Nucleic Acids Res. 2009, 37, 443–447.

There are several supplemental files that are not available in this version of the article. To view this additional information, please use the citation on the first page of this chapter.

CHAPTER 9

Detection of 140 Clinically Relevant Antibiotic-Resistance Genes in the Plasmid Metagenome of Wastewater Treatment Plant Bacteria Showing Reduced Susceptibility to Selected Antibiotics

RAFAEL SZCZEPANOWSKI, BURKHARD LINKE, IRENE KRAHN, KARL-HEINZ GARTEMANN, TIM GÜTZKOW, WOLFGANG EICHLER, ALFRED PÜHLER, AND ANDREAS SCHLÜTER

9.1 INTRODUCTION

Development and dissemination of antibiotic-resistance genes is a serious problem in the treatment of infectious diseases (Goossens, 2005; Lim & Webb, 2005). An important step in coping with this threat is to elucidate and to understand pathways for resistance gene spread. Many resistance genes are located on mobile genetic elements such as plasmids, transposons and integrons, which function as vectors for these determinants and promote their dissemination (Bennett, 1999; Davies, 1994; Davison, 1999; Hall & Collis, 1995; Mazel & Davies, 1999; Rowe-Magnus & Mazel, 1999; Seveno et al., 2002). Moreover, inappropriate use of antimicrobial drugs favours spread of resistance genes by selection for resistant micro-organisms (Bywater, 2004, 2005; Wassenaar, 2005).

Antibiotic-resistant bacteria of wastewater treatment plants (WWTPs) are the focus of the present study. WWTPs are connected to private house-holds and hospitals where antibiotics are used and resistances in bacteria might arise. Once antibiotic-resistant bacteria reach WWTPs, they poten-tially can disseminate their resistance freight among members of the endog-enous microbial community. Evidence for horizontal transfer of resistance elements in sewage habitats has been obtained for model systems (Geisen-berger et al., 1999; Marcinek et al., 1998; Nüßlein et al., 1992). Because of the favourable growth conditions they provide for many micro-organisms, WWTPs have to be considered as hot-spots for horizontal transfer of genetic material, e.g. by means of conjugation (Mach & Grimes, 1982; Mancini et al., 1987). In addition, contamination of sewage with antibiotics might cause a selective advantage for resistant bacteria (Göbel et al., 2005; Golet et al., 2002, 2003; Jarnheimer et al., 2004; Kümmerer, 2003; Kümmerer et al., 2000; Lee et al., 2007; Lindberg et al., 2005, 2006).

Previously, 12 different resistance plasmids, namely pB2/pB3 (Heuer et al., 2004), pB4 (Tauch et al., 2003), pB8 (Schlüter et al., 2005), pB10 (Schlüter et al., 2003), pTB11 (Tennstedt et al., 2005), pRSB101 (Szcz-epanowski et al., 2004), pRSB105 (Schlüter et al., 2007a), pRSB107 (Szczepanowski et al., 2005), pRSB111 (Szczepanowski et al., 2007), pGNB1 (Schlüter et al., 2007b) and pGNB2 (Bönemann et al., 2006), were isolated from WWTP compartments and analysed at the genomic and functional level. These plasmids confer resistance to different antibi-otics such as aminoglycosides, β-lactams, chloramphenicol, macrolides, quinolones, fluoroquinolones, tetracycline, trimethoprim and sulphon-amides. In addition, some of the plasmids analysed carry heavy metal, quaternary ammonium compound or triphenylmethane dye resistance genes. Moreover, different class 1 integron-specific resistance gene cassettes were identified on plasmids from WWTP bacteria (Tennstedt et al., 2003). A total of 22 different resistance genes and 27 different integron-specific resistance gene cassettes were identified on plasmids harboured by bacteria of activated sludge and the WWTP's final efflu-ents. Other studies investigated the occurrence of resistance genes in dif-ferent aquatic systems including sewage habitats. Many of these studies focused either on selected antibiotic-resistance genes, e.g. vanC, ampC, mecA (Schwartz et al., 2003; Volkmann et al., 2004), or on genes conferring resistance to a specific class of antimicrobial compounds, e.g. β-lastams (Henriques et al., 2006a, b), chloramphenicol (Dang et al., 2008), or tetracyclines (Akinbowale et al., 2007; Chee-Sanford et al., 2001; Guillaume et al., 2000; Smith et al., 2004).

A more comprehensive study investigated the plasmid metagenome of WWTP bacteria with reduced susceptibility to certain antimicrobial drugs by applying the next-generation 454-pyrosequencing technology (Schlüter et al., 2008; Szczepanowski et al., 2008). This approach led to the identification of sequences that are very similar to 81 different antibiotic-resistance genes, three multidrug efflux genes and three quaternary ammonium compound resistance genes. However, detailed analysis of the plasmid metagenome dataset indicated that the corresponding sequencing approach was not carried out to saturation. Thus, it is very likely that low-abundance genes were not detected. Moreover, only one compartment of the wastewater treatment plant was investigated by the cited plasmid metagenome study.

Therefore, the present study was aimed at screening the same WWTP for the occurrence of a large set of known antibiotic-resistance genes by means of a PCR approach which should also allow for detection of low-abundance resistance genes. The identification of resistance genes involved design and testing of 192 resistance-gene-specific PCR primer pairs. The question of whether the set of resistance determinants could also be detected in the WWTP's final effluents was also addressed.

9.2 METHODS

9.2.1 ISOLATION OF PLASMIDS FROM RESISTANT BACTERIA RESIDING IN ACTIVATED SLUDGE AND THE FINAL EFFLUENTS OF THE WWTP

The WWTP samples were taken in September 2006 from the municipal WWTP Bielefeld-Heepen, Germany. One litre of the final effluent sample was centrifuged (5 min, 8000 g) and the resulting pellet was resuspended in 5 ml Luria Broth. Aliquots (100 µl) of the resuspended final effluent sample and the activated sludge sample were plated in five replicates in serial dilutions onto Luria-Broth agar plates supplemented with one of the following antibiotics: 100 μg ampicillin ml^{-1}, 1 μg cefotaxime ml^{-1}, 15 μg cefuroxime ml^{-1}, 25 μg chloramphenicol ml^{-1}, 1 μg ciprofloxacin ml^{-1}, 200 μg erythromycin ml^{-1}, 15 μg gentamicin ml^{-1}, 50 μg kanamycin ml^{-1}, 1 μg norfloxacin ml^{-1}, 30 μg rifampicin ml^{-1}, 100 μg spectinomycin ml^{-1}, 100 μg streptomycin ml^{-1}, 5 μg tetracycline ml^{-1}. The agar medium was also supplemented with cycloheximide at a final concentration of 75 μg ml^{-1} to avoid growth of fungi. After incubation at 30 °C for 36 h the bacteria were collected separately for each antibiotic used for selection. Total plasmid DNAs from activated sludge or final effluent bacteria were prepared with the NucleoBond kit PC100 on AX 100 columns (Macherey-Nagel) according to the manufacturer's protocol. This method has been shown to be suitable for isolation of plasmids in a size range of 40 to 180 kbp (Stiens et al., 2008; Szczepanowski et al., 2004), with the limitation that larger plasmids cannot be isolated with the same efficiency as smaller plasmids. It should also be mentioned here that the plasmid isolation procedure is biased by the lysis method implemented in the NucleoBond kit PC100 protocol since it cannot be assumed that all kinds of WWTP bacteria are equally well lysed by this method. After DNA isolation, a CsCl high-density gradient centrifugation (Sambrook et al., 1989) using a Vti 65.2 rotor was performed in order to minimize contamination with chromosomal DNA. Plasmid DNA concentrations were determined by using the NanoDrop 1000 instrument (NanoDrop Technologies). For further analyses, 20 μl of each total plasmid DNA preparation (separately held for plasmid DNAs from activated sludge and final effluent bacteria) were mixed, resulting in two master total plasmid DNA samples.

9.2.2 SELECTION OF TARGET REFERENCE ANTIBIOTIC-RESISTANCE GENES AND DESIGN OF SPECIFIC PCR PRIMERS

For the design of resistance-gene-specific PCR primers, reference resistance gene nucleotide sequences were extracted from different databases: EBI SRS server (http://srs.ebi.ac.uk/), NCBI (http://www.ncbi.nlm.nih.gov/), β-lactamase genes (http://www.lahey.org/Studies/) and macrolide and tetracycline resistance genes (http://faculty.washington. edu/marilynr/). In total, about 650 resistance and multidrug efflux permease gene sequences known to confer resistance to different antimicrobial compounds including aminoglycosides, β-lactams, chloramphenicol, macrolides, quinolones, fluoroquinolones, rifampicin, tetracyclines, trimethoprim, sulphonamides and quaternary ammonium compounds were selected from these databases. A new database, named ARG-DB (Antibiotic Resistance Gene Database), was set up for the extracted genes and all entries of ARG-DB were compared to each other by applying the BLAST algorithm. Genes with more than 85% sequence identity were clustered. Based on CLUSTAL W (Larkin et al., 2007) alignments, a consensus sequence was calculated for each cluster and the gene showing the highest degree of identity to the consensus sequence was defined as representative for the respective cluster. This approach led to the selection of 192 reference resistance genes (see Table 1 and Supplementary Table S1, available with the online version of this paper) each representing a distinct alignment cluster. Specific PCR primers were designed for all reference genes by means of the Primer3 program (Rozen & Skaletsky, 2000) and synthesized. The resulting PCR primer sequences are shown in Table S1. Plasmid incompatibility genes specific for the Inc groups P, Q, W, N (Götz et al., 1996), A/C (Llanes et al., 1996) and F (Eichenlaub et al., 1977) as well as the genes gfp (Prasher et al., 1992) and luc (accession no. D25416) were chosen as control sequences for primer design.

9.2.3 PCR AND AMPLICON DETECTION

The reaction mix of the PCR was composed of approximately 100 ng total plasmid DNA as template, 2.5 µl reaction buffer (10×), 2 mM $MgCl_2$, 0.2 mM of each dNTP, 0.5 µM of each primer, 1 U Taq DNA polymerase (BioLine), and filled up to 25 µl with sterile double-distilled water. The initial step of the reaction was denaturation of DNA at 94 °C for 4 min. This step was followed by 35 cycles composed of 1 min denaturation at 94 °C, 1 min annealing at 58 °C and 45 s polymerization at 72 °C. The final polymerization step was performed for 10 min at 72 °C. The amplicons were analysed by gel electrophoresis (in 1% agarose in Tris/

TABLE 1. Taxon composition profile of tetracycline resistant bacteria (TRB) in activated sludge.

Antimicrobial compound class	Encoded enzymes	Gene names
Aminoglycosides	Aminoglycoside acetyltransferases	aacA, aacA1, aacA4, aacA7, aacA29b, aacC1, aacC2, aacC3, aacC4, aac(3)-Id, aac(6′)-Im
	Aminoglycoside adenylyltransferases	aadA4, aadA7, aadA9, aadA10, aadA12, aadD
	Aminoglycoside phosphotransferases	aph, aphA, aphA-3, aphA-6, aphA-7, aph2, aph(2′)-Ib, strA, strB
β-Lactams	Class A β-lactamases	tx-m4, ctx-m26, ctx-m27, ctx-m32, ges-3, kpc-3, per-1, per-2, shv-34, bla_{TEM-1}, bla_{TLA-1}, bla_{TLA-2}, veb-1
	Class B β-lactamases	imp-2, imp-5, imp-9, imp-13, imp-16, imp-16, vim-4, vim-7
	Class C β-lactamases	ampC, cmy-9, cmy-13
	Class D β-lactamases	bla_{nps-1}, bla_{nps-2}, oxa-1, oxa-2, oxa-5, oxa-9, oxa-10, oxa-12, oxa-18, oxa-20, oxa-22, oxa-27, oxa-29, oxa-40, oxa-45, oxa-46, oxa-48, oxa-50, oxa- 54, oxa-55, oxa-58, oxa-60, oxa-61, oxa-75, mecA
Chloramphenicol/florfenicol	Chloramphenicol acetyltransferases	cat, cat, cat, cat, cat2, catIII, catA, catB2, catB4, catB6, catB7, catB8, catB9, catP, cat-TC
	Chloramphenicol/florfenicol transporters	cmlA1, cmxA, fexA, floR
	Hydrophobic polypeptide	cmlB
Fluoroquinolones	Pentapeptide family proteins	qnrA3, qnrB1, qnrB4, qnr
Macrolides	rRNA adenine N^6-methyltransferases	ermA, ermB, ermD, ermF, erm(A), erm(TR)
	Esterase	ereA2, ereB
	MFS efflux proteins	mefA, mefE, mefE, mel, msr(A)
	Macrolide 2′-phosphotransferases	mph(B), mph(A), mph, mphB, mph(BM)
	Hydrolase	vgh(A)
	Streptogramin B lactonase	vgbB
Rifampicin	ADP-ribosylating transferase	arr2

TABLE 1. CONTINUED.

Tetracyclines	Tetracycline transporters	tet(A), tet(A), tetA(C), tetA(E), tetA(J), tetBSR, tet(D), tet(G), tet(H), tet(L), tetA(Y), tet(Z), effJ, tet(V), tet(K), tet(30), tet(33), tet(38), tetA(39)
	Tetracycline inactivation proteins	tet(37), tet(X)
	GTP-binding elongation factor proteins	tetB(P), tet(M), tet(M), tet(M), tet(M), tet(O), tet(S), tet(W), tet(32)
	Ribosomal protection tetracycline resistance proteins	tet(36), tetQ, tet(T)
	Tetracycline repressor protein	tetR(31)
	Tetracycline resistance	tet(U)
	Phosphoribosyltransferase	tet(34)
Trimethoprim	Dihydrofolate reductases	dfrII, dfrV, dfrVI, dfrXII, dfr13, dfr16, dfr17, dfrA19, dfrB2, dfrD, dhfr, dhfR, dhfrI, dhfrVIII, dhfrIX, dhfrXV
Sulfonamides	Dihydropteroate synthetases	sulI, sulII, sulIII
Quaternary ammonium compounds	Small multidrug efflux proteins	qacB, qacD, qacEΔ1, qacF, qacF, qacG, qacG2, qacH
Various antibiotics transported by multidrug efflux genes	Multidrug efflux pumps	acrB, acrD, mexB, mexD, mexD, mexF, mexI, mexY, orfl1
Total		192

HCl/acetate buffer), stained with ethidium bromide and visualized under UV light.

9.2.4 SEQUENCING AND ANALYSIS OF SELECTED RESISTANCE-GENE-SPECIFIC AMPLICONS

After filter purification by means of MAHVN 4550 (Millipore) and G-50 Fine Sephadex (Sigma-Aldrich) the amplicons were sequenced on an ABI

3730 XL sequencer (Applera, Applied Biosystems) using Big Dye 3.1 chemistry. Assembly of the forward and reverse sequence of each amplicon, and sequence quality control, was carried out by means of the CONSED/AUTO-FINISH software tool (Gordon et al., 1998, 2001). Assembled resistance-gene-specific amplicon sequences were compared to the NCBI nucleotide sequence database by means of BLAST (Altschul et al., 1990).

9.3 RESULTS AND DISCUSSION

9.3.1 ISOLATION OF ANTIBIOTIC-RESISTANCE PLASMIDS FROM RESISTANT BACTERIA OBTAINED FROM ACTIVATED SLUDGE AND THE WWTP'S FINAL EFFLUENTS

To get an overview of the occurrence of resistance determinants in a WWTP habitat, total plasmid DNA preparations isolated from antibiotic-resistant WWTP bacteria were probed for different known resistance genes by means of a PCR approach. Antibiotic-resistant bacteria originating from activated sludge or from the final effluent compartment of the municipal WWTP Bielefeld-Heepen were selected on media supplemented with one of 12 clinically relevant antibiotics (see Methods). Total plasmid DNA was prepared from bacteria able to grow on these selective media and used as template in PCR analyses for the detection of selected resistance determinants. The concentration of the pooled template DNAs was about 80 ngµl^{-1} for each habitat (activated sludge and final effluents). Target reference resistance genes were extracted from different nucleotide sequence databases, and gene-specific PCR primers were designed (see Supplementary Table S1).

9.3.2 DETECTION OF PLASMID-ENCODED RESISTANCE GENES IN RESISTANT BACTERIA ISOLATED FROM ACTIVATED SLUDGE AND THE WWTP'S FINAL EFFLUENTS

To detect resistance genes and plasmid incompatibility determinants present in bacteria residing in activated sludge and the final effluent compartment of the WWTP, PCR analyses using 200 specific primer pairs were carried out. Total plasmid DNA preparations from antibiotic-resistant WWTP bacteria were used as template DNAs in these PCRs. In total, 145 amplicons (140 specific for resistance genes and five for plasmid incompatibility determinants) were obtained in these PCRs on total plasmid DNA from antibiotic-resistant activated-sludge bacteria (Table 2). The total plasmid DNA preparation originating from bacteria of the final-effluent compartment yielded 129 amplicons (123 specific for resistance genes and six for plasmid-specific genes) (Table 2). PCR results were positive for resistance genes known to confer resistance to different aminoglycoside, β-lactam, chloramphenicol, fluoroquinolone, macrolide, rifampicin, tetracycline, trimethoprim and sulfonamide antibiotics as well as to quaternary ammonium compounds (Table 2).

Results of this study were compared to the plasmid metagenome data that were recently obtained for activated sludge bacteria showing reduced susceptibility to selected antimicrobial drugs from the same WWTP. High-throughput 454-pyrosequencing of plasmids from these bacteria revealed that numerous sequences are very similar or even identical to 81 known antibiotic-resistance genes conferring resistance to the major classes of antimicrobial drugs (Schlüter et al., 2008; Szczepanowski et al., 2008). The PCR-based approach led to the detection of 59 additional resistance genes in activated-sludge bacteria that were not apparent in the plasmid metagenome dataset. For instance, 15 additional tetracycline resistance genes appeared in the PCR analysis. In contrast, only sequences for seven different tetracycline-resistance genes, namely tetA(A), tetA(B), tetA(C), tetA(D), tetA(E), tetA(X) and tet(39), were identified in the metagenome dataset (Schlüter et al., 2008; Szczepanowski et al., 2008). Moreover, the PCR-based approach led to the detection of 123 different plasmid-encoded resistance-gene-specific amplicons in bacteria isolated from the final effluent of the WWTP analysed here. This compartment was not covered by the cited plasmid metagenome study.

The present study also showed that the resistance gene spectra detected in plasmid DNA preparations originating from activated-sludge and from final-effluent bacteria are quite similar. Only the numbers of detected aminoglycoside, β-lactam, macrolide and tetracycline resistance genes differ slightly for the WWTP compartments tested.

TABLE 2. Taxon composition profile of tetracycline resistant bacteria (TRB) in activated sludge.

Gene name*	Gene product	Amplicon size (bp)	Resistance to/func-tion†	Detected in activated sludge	Detected in the final effluents	Accession no.
aacA, aadB	Aminoglycoside 6′-N-acetyl-transferase	197	Km, Tob, Ak	+	+	M86913
aacA1	Aminoglycoside 6′-N-acetyl-transferase	200	Gm, Km, Tob, Neo	+	–	AB113580
aacA4	Aminoglycoside 6′-acetyl-transferase	196	Ak	+	+	AJ744860
aacA7	Aminoglycoside acetyltrans-ferase-6′ type I	175	Gm, Tob, Km	+	–	AF263520
aacA29b	Aminoglycoside 6′-N-acetyl-transferase	170	Ak, Km	+	+	AY139599
aacC1	Aminoglycoside 3N-acetyl-transferase	130	Gm	+	+	AY139604
aacC2	Aminoglycoside (3)-N-acetyltransferase	148	Gm	+	+	S68058
aacC4	Aminoglycoside (3)-acetyl-transferase IV	147	Gm	+	+	X01385
aac(3)-Id‡	3′-N-Aminoglycoside acetyl-transferase	178	Gm	+	+	AY458224
aac(6′)-Im	6′-Aminoglycoside N-acetyltransferase	194	Tob, Ak, Km	+	+	AF337947
aadA4, aadA5	Streptomycin 3′-adenylyl-transferase	198	Sm, Sp	+	+	AY138986

TABLE 2. CONTINUED.

aadA7	Aminoglycoside (3')(9)-adenylyltransferase	187	Sm, Sp	+	+	AY463797
aadA9	Streptomycin 3'-adenylyltransferase	184	Sm, Sp	+	−	AJ420072
aadA10, aadA6/aadA10‡	Aminoglycoside (3')(9)-adenylyltransferase	198	Sm, Sp	+	+	U37105
aadA12, aadA1, aadA2, aadA8, aadA11, aadA13, aadA23	Putative streptomycin 3'-adenylyltransferase	186	Putative Sm, Sp	+	+	AY665771
aadD	Kanamycin-nucleotidyltransferase	153	Km	+	+	AB037420
aph	Aminoglycoside 3'-phosphotransferase	173	Km, Neo	+	+	AJ851089
aphA	3'-Aminoglycoside phosphotransferase	198	Km	+	+	AJ744860
aphA-3	3'5'-Aminoglycoside phosphotransferase of type III	139	Km	+	+	V01547
aphA-6	3'-Aminoglycoside phosphotransferase	192	Km, Ak	+	+	X07753
aph2	Aminoglycoside-3'-O-phosphotransferase	198	Km, Neo	+	+	U00004
aph(2')-Ib	Aminoglycoside phosphotransferase	175	Km	+	−	AF337947
strA	Aminoglycoside 3'-phosphotransferase	196	Sm	+	+	NC_004840

TABLE 2. CONTINUED.

strB	Aminoglycoside 6-phosphotransferase	150		+	+	NC_004840
ctx-m-4	Class A β-lactamase	155	Amp, Ctx, Cxm, Atm	+	+	Y14156
ctx-m-27‡	Class A β-lactamase	158	Caz, Ctx, Amo, Tic, Prl, Kf, Cxm, Cpo, Atm	+	+	AY156923
ctx-m-32‡	Class A β-lactamase	156	Amo, Ctx, Caz, Fep, Prl, Kf, Fox, Cxm	+	+	AJ557142
ges-3‡	Class A extended-spectrum β-lactamase	181	Titeracillin, Prl, Caz, Ctx, Atm, Ipm	+	+	AY494717
per-2	Class A extended-spectrum β-lactamase	198	Oxyiminocephalosporins, Atm, Cft	+	−	X93314
shv-34	Class A β-lactamase	200	Caz, Ctx	+	+	AY036620
bla$_{TEM-I}$	Class A β-lactamase	167	Amp, Pen-G	+	+	AJ851089
bla$_{TLA-2}$	Class A extended spectrum β-lactamase	186	Amo, Tic, Caz, Kf, Cxm, Fox, Ctx, Fep, Atm	+	+	NC_006385
veb-1	Class A extended-spectrum β-lactamase	190	Cephalosporins, Atm	+	−	AF010416
vim-4	Metallo-β-lactamase	171	β-Lactams	−	+	AY509609
imp-2, imp-5	Class B metallo β-lactamase	200	Amp, Ctx, Fep	+	+	AJ243491
imp-9‡, imp-11	Class B metallo β-lactamase	178	β-Lactams	+	+	AY033653
imp-13, imp-2	Class B metallo β-lactamase	198	Cxm, Caz, Ctx, Cro, Fep, Amp	+	+	AJ550807

TABLE 2. CONTINUED.

ampC	Class C β-lactamase, cepha-losporinase	189	Pen, cephalosporins	+	+	J01611
cmy-9, cmy-10	Class C β-lactamase	169	β-Lactams	+	+	AB061794
cmy-13‡, cmy-5	Class C β-lactamase	150	β-Lactams	+	+	AY339625
bla$_{NPS-1}$	Class D β-lactamase	188	Amo, azlocillin, Cec, cefazolin, Cfp, Prl	–	–	NC_003430
bla$_{NPS-2}$	Class D β-lactamase	192	Amp	+	+	NC_006388
oxa-1	Class D β-lactamase	199	β-Lactams	+	+	AY139600
oxa-2, oxa-21, oxa-53	Class D β-lactamase	177	β-Lactams	+	+	NC_007502
oxa-5	Class D β-lactamase	175	β-Lactams	+	+	X58272
oxa-9	Class D β-lactamase	162	β-Lactams	+	–	M55547
oxa-10, oxa-56	Class D β-lactamase	191	β-Lactams	+	+	AY115475
oxa-12	Class D β-lactamase	188	β-Lactams	+	+	U10251
oxa-20	Class D β-lactamase	163	Amo, Tic	–	+	AF024602
oxa-22	Class D β-lactamase	200	Benzylpenicillin, Ob	+	+	AF064820
oxa-27	Class D β-lactamase	180	β-Lactams	+	–	AF201828
oxa-40	Class D β-lactamase	168	Amo, Tic, Caz, Fep, Cpo, Prl, Kf, Cxm, Ipm	+	+	AF509241
oxa-46, oxa	Class D β-lactamase	150	Amp, Car, Mez, Kf	+	+	AF317511
oxa-48	Class D β-lactamase	145	Amo, Tic, Fep, Ipm, Cpo, Prl, Ctx	+	+	AY236073

TABLE 2. CONTINUED.

oxa-50	Class D β-lactamase	198	Amp, Tic, Ctx, Prl, Kf, Cxm	+	+	AY306130
oxa-58‡	Class D β-lactamase	152	Amo, Tic, Cpo, Prl, Ipm, Kf	+	+	AY665723
oxa-75	Class D β-lactamase	181	Amp, Prl	+	+	AY859529
cmlA1, cmlA5	Chloramphenicol efflux protein	137	Cm	+	+	NC_006388
cmlB	Hydrophobic polypeptide	147	Cm	+	+	AF034958
cmxA	Chloramphenicol export protein	186	Cm	+	+	AF024666
fexA	Florfenicol/chloramphenicol exporter	198	Cm, Ffc	+	–	AJ549214
floR, cmlA	Efflux protein	188	Cm, Ffc	+	+	AF118107
cat	Chloramphenicol acetyltransferase	173	Cm	+	+	M11587
cat	Chloramphenicol acetyltransferase	162	Cm	+	+	M35190
cat	Chloramphenicol acetyltransferase	195	Cm	+	+	S48276
cat	Chloramphenicol acetyltransferase	163	Cm	+	+	M58515
cat2, catII, cmlA	Chloramphenicol acetyltransferase	192	Cm	+	+	AY509004
catII	Chloramphenicol acetyltransferase	150	Cm	+	+	X07848

TABLE 2. CONTINUED.

catA	Chloramphenicol acetyltrans-ferase	186	Cm	+	+	AJ851089
catB2	Chloramphenicol acetyltrans-ferase	156	Cm	+	+	AY139601
catB4	Chloramphenicol acetyltrans-ferase	188	Cm	+	+	AF322577
catB6	Chloramphenicol acetyltrans-ferase	144	Cm	–	+	AJ223604
catB7	Chloramphenicol acetyltrans-ferase	152	Cm	+	+	AF036933
catB8	Chloramphenicol acetyltrans-ferase	175	Cm	+	+	AF227506
cat-TC, cat	Chloramphenicol acetyltrans-ferase	194	Cm	+	+	U75299
qnrA3‡, qnr	Pentapeptide family, DNA-gyrase and topoisomerase IV protection	168	Nal	+	+	DQ058661
qnrB1‡, qnrB2, qnrB5	Pentapeptide family, DNA-gyrase and topoisomerase IV protection	191	Cip	+	+	DQ351241
qnrB4	Pentapeptide family	158	Quinolones	+	+	DQ303921
qnr, qnrS2‡	Quinolone resistance deter-minant	175	Cip, Nor, Nal	+	+	AB187515
ereA2, ereA	Erythromycin esterase type I	177	Em	+	+	AF512546
ereB	Erythromycin esterase type II	158	Em	+	–	X03988

TABLE 2. CONTINUED.

mph(B)	Macrolide phosphotransferase	199	Azi, Cla, Em, Rox, Tyl	+	+	AM260957
mph(A)	Macrolide 2′-phosphotransferase I	153	Azi, Cla, Em, Rox	+	+	NC_006385
mph	Macrolide 2′-phosphostransferase	200	Em	+	+	DQ839391
mph(B)	Macrolide 2′-phosphotransferase II	200	Macrolides	+	+	D85892
mphBM	Macrolide 2′-phosphotransferase II	200	Macrolides	+	–	AF167161
ermA	rRNA adenine N^6-methyltransferase	185	Em	+	–	X51472
ermB	rRNA adenine N^6-methyltransferase	193	Em	+	+	M11180
ermF	rRNA adenine N^6-methyltransferase	323	MLS	+	+	M14730
mef(A)	Macrolide-efflux protein, MFS permease	179	Em	+	–	AJ715499
mefE, mefI	Macrolide-efflux protein, MFS permease	199	Em	+	–	AF274302
mel	Macrolide-efflux protein, macrolide-specific ABC-type efflux carrier	198	Azi, Cla, Em	+	+	DQ839391
msrA	Erythromycin resistance ATP-binding protein MsrA	158	Em	+	–	X52085

TABLE 2. CONTINUED.

			Rif			
arr2	Putative rifampicin ADP-ribosyltransferase	140	Rif	+	+	AF205943
sulI	Dihydropteroate synthetase	185	Sul	+	+	NC_006388
sulII	Dihydropteroate synthetase	147	Sul	+	+	AJ851089
sul3	Dihydropteroate synthetase	199	Sul	+	+	AY316203
dfrII	Dihydrofolate reductase	156	Tp	+	+	AY139601
dfrV	Dihydrofolate reductase	180	Tp	+	+	AY139589
dfr13(dfrXIII)	Dihydrofolate reductase	174	Tp	+	+	Z50802
dfr16	Dihydrofolate reductase	173	Tp	+	+	AY259085
dfr17, dfrVII	Dihydrofolate reductase	152	Tp	+	+	AY139588
dfrA19	Dihydrofolate reductase	165	Tp	+	+	AM234698
dfrB2	Dihydrofolate reductase	198	Tp	+	+	AY139592
dfrD	Dihydrofolate reductase	194	Tp	+	+	Z50141
dhfr1	Dihydrofolate reductase	169	Tp	+	+	AJ698325
dhfrVIII	Dihydrofolate reductase	169	Tp	+	+	U10186
dhfrXV	Dihydrofolate reductase	197	Tp	+	+	Z83311
tetA	MFS tetracycline efflux	200	Tc	+	+	NC_004840
tetA	MFS tetracycline efflux	198	Tc	+	+	NC_006388
tetA	MFS tetracycline efflux	187	Tc	+	+	AJ851089
tetA	MFS tetracycline efflux	176	Tc	+	+	L06940
tetD	MFS tetracycline efflux	155	Tc	+	+	L06798
tetG	MFS tetracycline efflux	140	Tc	+	+	AF133139
tetH	MFS tetracycline efflux	164	Tc	+	+	AJ245947

TABLE 2. CONTINUED.

tetL	MFS tetracycline efflux	176	Tc	+	–	U17153
tet(U)	Replication	198	Low level Tc	+	+	U01917
tetY	MFS tetracycline efflux	146	Tc	+	+	AF070999
tetR(31)	Tetracycline repressor protein	168	Regulates expression of TetA(31)	+	+	AJ250203
effJ (tet(35))	Putative tetracycline efflux pump	190	Tc	–	+	AF35362
tet(39)	MFS tetracycline efflux	154	Tc	+	+	AY743590
tetB(P)	GTP-binding elongation factor protein, TetM/TetO family	143	Tc	+	–	L20800
tet(M)	GTP-binding elongation factor protein, TetM/TetO family	197	Tc	+	+	M21136
tet(M)	GTP-binding elongation factor protein, TetM/TetO family	197	Tc	+	+	M85225
tet(M)	GTP-binding elongation factor protein, TetM/TetO family	198	Tc	+	+	X04388
tet(M)	GTP-binding elongation factor protein, TetM/TetO family	198	Tc	+	+	X90939
tet(O)	GTP-binding elongation factor protein, TetM/TetO family	189	Tc	+	–	Y07780

TABLE 2. CONTINUED.

Gene	Description	No.	Type			Accession
tet(S)	GTP-binding elongation factor protein, TetM/TetO family	172	Tc	+	+	L09756
tet(32)	GTP-binding elongation factor protein, TetM/TetO family	149	Tc	+	–	AJ295238
tet(36)	Ribosomal protection tetra-cycline resistance protein	192	Tc	+	+	AJ514254
tet(X)	Inactivation of tetracycline	186	Tc	–	+	M37699
qacB	Permease of the MFS family, multidrug efflux protein	164	Multidrug efflux	+	–	AF053771
qacEΔ1	Small multidrug resistance protein, membrane trans-porter of cations and cationic drugs	198	QAC	+	+	AJ698325
qacF	Small multidrug resistance protein, membrane trans-porter of cations and cationic drugs	195	QAC	+	–	NC_007502
qacF, qacH	Small multidrug resistance protein, membrane trans-porter of cations and cationic drugs	172	QAC	+	+	AY139598
qacG2	Small multidrug resistance protein, membrane trans-porter of cations and cationic drugs	147	QAC	+	+	AJ609296

TABLE 2. CONTINUED.

acrB	RND family, acridine/multi-drug efflux pump	160	Multidrug efflux	+	+	M94248
acrD	Cation/multidrug efflux pump	185	Aminoglycosides, Nv	+	+	U12598
mexB	Cation/multidrug efflux pump, RND multidrug efflux transporter	147	Multidrug efflux	+	+	L11616
mexD	RND multidrug efflux transporter	185	Em, Rox	+	+	NC_003430
mexD	Cation/multidrug efflux pump, RND multidrug efflux transporter	182	Multidrug efflux	+	+	U57969
mexF	Cation/multidrug efflux pump, RND multidrug efflux transporter	348	Multidrug efflux	+	+	X99514
mexI	Cation/multidrug efflux pump, RND multidrug efflux transporter	170	Multidrug efflux	+	+	AE004837
mexY	Cation/multidrug efflux pump, RND multidrug efflux transporter	198	Multidrug efflux	+	+	AB015853
orf11	ABC type permease	198	Nal, Nor	+	+	NC_006385
kikA	Killing in Klebsiella	198	IncN-specific gene	+	+	AY046276
oriV	Origin of vegetative replication	171	IncW-specific region	−	+	BR000038

TABLE 2. CONTINUED.

oriV	Origin of vegetative replica-tion	192	IncQ-specific region	+	+	NC_001740
rep	Replication initiation protein	163	IncA/C-specific gene	+	+	X73674
repE	Replication initiation protein	192	IncFIA-specific replication gene	+	+	AJ851089
trfA	Replication initiation protein	192	Initiation of replica-tion, IncP-specific gene	+	+	NC_004840

*The PCR product is specific for all genes given in the field.

†Resistance spectra data were extracted from the respective database entry and the literature cited therein. Abbreviations: Ak, amikacin; Amo, amoxicillin; Amp, ampicillin; Atm, aztreonam; Azm, azithromycin; Car, carbenicillin; Caz, ceftazidim; Cec, cefaclor; Cfp, cefoperazon; Cft, ceftibuten; Cip, ciprofloxacin; Clr, clarithromycin; Cm, chloramphenicol; Cpo, cefpirom; Cro, ceftriaxon; Ctx, cefotaxime; Cxm, cefuroxime; Em, erythromycin; Fep, cefepim; Ffc, florfenicol; Fox, cefoxitin; Gm, gentamicin; Ipm, imipenem; Kf, cephalothin; Km, kanamycin; Lev, levofloxacin; Met, meticillin; MLS, macrolide-lincosamide-streptogramin B; Mez, mezlocillin; Nal, nalidixic acid; Neo, neomycin; Nor, norfloxacin; Nv, novobiocin; Ob, cloxacillin; Ofx, ofloxacin; Pen-G, penicillin G; Prl, piperacillin; QAC, quaternary ammonium compounds; Rif, rifampicin; Rox, roxithromycin; Spar, sparfloxacin; Sm, streptomycin; Sp, spectinomycin ; Sul, sulfonamides; Tc, tetracyclines; Tic, ticarcillin; Tob, tobramycin; Tp, trimethoprim; Ty, tylosin.

‡Resistance genes recently described in clinical isolates.

Interestingly, the same fluoroquinolone, trimethoprim and sulfonamide resistance genes as well as the same genes for multidrug efflux systems could be detected in both plasmid samples (see Table 2). This high congruence of amplicons for the latter resistance genes may be explained by the fact that antibiotics, especially fluoroquinolones, trimethoprim and sulfonamides, are only poorly removed during wastewater treatment processes (Göbel et al., 2005; Golet et al., 2002, 2003; Lindberg et al., 2005, 2006; Nakata et al., 2005) and therefore might exert selective pressure on bacteria within the sewer system or the sewage plant, leading to enrichment of resistant bacteria and their release into the environment with the final effluents.

9.3.3 DETECTION OF RESISTANCE GENES RECENTLY DESCRIBED FROM CLINICAL ISOLATES IN THE WWTP COMPARTMENTS ANALYSED

Detection of numerous and various resistance genes in bacteria from activated sludge (140 genes) and the final effluents of the WWTP (123 genes) raises the question whether resistance genes recently described for clinical isolates are also present in and are released from the municipal sewage plant under study. It appeared that the aminoglycoside resistance genes aadA6/aadA10 (Fiett et al., 2006) and aac(3)-Id (Doublet et al., 2004), the β-lactam resistance genes ctx-m-27 (Bonnet et al., 2003), ctx-m-32 (Cartelle et al., 2004), ges-3 (Vourli et al., 2004), imp-9 (Xiong et al., 2006), imp-13 (Toleman et al., 2003) and oxa-58 (Poirel et al., 2005), and the fluoroquinolone resistance genes qnrA3 (Heritier et al., 2004), qnrB1 (Jacoby et al., 2006) and qnrS (Hata et al., 2005), which were recently described as new genes or novel variants of known genes in clinical isolates, could be identified in the WWTP analysed (Table 2).

9.3.4 SEQUENCING OF SELECTED RESISTANCE-GENE-SPECIFIC AMPLICONS TO VERIFY THEIR IDENTITY

To verify the identity of the PCR products obtained in the analyses described above, 45 amplicons were randomly selected and sequenced. Sequencing and annotation results are summarized in Table 3. The nucleotide sequences of 20 amplicons (bla$_{TLA-2}$, ctx-m27, ges-3, imp-13, oxa-58, veb-1, cmxA, qnr, qnrB4, ereA2, arr2, tetB(P), tetL, tet(M), tet(S), tet(X), acrD, mexD, mexI, mexY) are identical to the corresponding reference sequences, and eight amplicons (ampC, ctx-m-32, qnrB1, ermB, tetH, tet(O), acrB, mexD) display only one nucleotide exchange compared to the reference sequence. Moreover, the nucleotide sequences of 15 further amplicons are 87% to 98% identical to the corresponding reference gene. In the case of two amplicons, namely those for the genes tetA(39) and cmlB, only the sequence from one sequencing direction could be obtained. The resulting short sequence reads show, respectively, 100% identity (over a length of 99 bases) to tetA(39) and 93% identity (over a length of 101 bases) to cmlB. Although some amplicon sequences do not show the highest degree of identity to the corresponding reference resistance gene, they are very similar or even identical to a closely related resistance determinant. For example, the amplicon sequence obtained with primers designed on the resistance gene ctx-m-32 is identical to the sequence of ctx-m-64 and has only one mismatch compared to the reference gene ctx-m-32. These results show that the sequenced amplicons really contain resistance-gene-specific nucleotide sequences.

9.4 CONCLUSIONS

This comprehensive study provides evidence that bacteria residing in different compartments of the WWTP analysed harbour various plasmid-borne resistance determinants representing all common classes. To our best knowledge, this is the first study that describes detection of resistance genes known to confer resistance to all common classes of antibiotics in two different compartments of the same WWTP. The mobile pool of resistance genes shared by bacteria of the WWTP analysed even includes resistance genes that have only recently been described for clinical isolates, indicating genetic exchange between clinical and WWTP bacteria. Moreover, detection of these newer resistance genes on plasmids isolated

TABLE 3. Sequencing of randomly selected resistance-gene-specific amplicons obtained from wastewater treatment plant bacteria and annotation results.

Gene-specific amplicon	Gene accession no.	Identity*	DNA-sequence identity (%)	Best hit to	Best hit accession no.
ampC	J01611	188 bp/189 bp	99	ampC	J01611
bla$_{TLA-2}$	NC_006385	186 bp/186 bp	100	bla$_{TLA-2}$	NC_006385
cmy-13	AY339625	149 bp/150 bp	99	cmy-28	EF561644
		147 bp/150 bp	97	cmy-13	AY339625
ctx-m-4	Y14156	154 bp/155 bp	99	ctx-m2	EF592570
		152 bp/155 bp	96	ctx-m4	Y14156
ctx-m-32	AJ557142	156 bp/156 bp	100	ctx-m64	AB284167
		155 bp/156 bp	99	ctx-m32	AJ557142
ctx-m-27	AY156923	158 bp/158 bp	100	ctx-m27	AY156923
ges-3	AY494717	181 bp/181 bp	100	ges-3	AY494717
imp-9	AY033653	176 bp/178 bp	98	imp-13	AJ628135
		157 bp/178 bp	87	imp-9	AY033653
imp-13	AJ550807	198 bp/198 bp	100	imp-13	AJ550807
oxa-46	AF317511	144 bp/150 bp	96	oxa-46	AF317511
oxa-58	AY665723	152 bp/152 bp	100	oxa-58	AY665723
shv-34	AY036620	199 bp/200 bp	99	shv-77	EF373975
		197 bp/200 bp	98	shv-34	AY036620
veb-1	AF010416	190 bp/190 bp	100	veb-1	AF010416
cmlB	AF034958	94 bp/101 bp (one read with 101 bp)	93	cmlB	AF034958
cmxA	AF024666	186 bp/186 bp	100	cmxA	AF024666
floR	AF118107	186 bp/188 bp	98	floR	AF118107
qnr	AB187515	175 bp/175 bp	100	qnr	AB187515
qnrA3	DQ058661	166 bp/168 bp	98	qnr	AY675584
		162 bp/168 bp	96	qnrA3	DQ058661
qnrB1	DQ351241	191 bp/191 bp	100	qnrB2	AM234698
		190 bp/191 bp	99	qnrB1	DQ351241
qnrB4	DQ303921	158 bp/158 bp	100	qnrB4	DQ303921
ereA2	AF512546	177 bp/177 bp	100	ereA2	AF512546
ermB	M11180	192 bp/193 bp	99	ermB	M11180
ermF	M14730	319 bp/323 bp	98	ermF	M14730
mef(A)	AJ715499	173 bp/179 bp	96	mefA	AJ715499

mefE	AF274302	198 bp/199 bp	99	mef	DQ445269
		194 bp/199 bp	96	mefE	AF274302
arr2	AF205943	140 bp/140 bp	100	arr2	AF205943
tetA(39)	AY743590	99 bp/99 bp (one read with 99 bp)	100	tetA(39)	AY743590
tetB(P)	L20800	143 bp /143 bp	100	tetB(P)	L20800
tetD	L06798	153 bp/155 bp	98	tetD	L06798
tetG	AF133139	138 bp/140 bp	98	tetG	AF133140
		132 bp/140 bp	94	tetG	AF133139
tetH	AJ245947	163 bp/164 bp	99	tetH	AJ245947
tetL	U17153	176 bp/176 bp	100	tetL	U17153
tet(M)	X90939	198 bp/198 bp	100	tet(M)	EF101931
		194 bp/198 bp	97	tet(M)	X90939
tet(M)	M21136	197 bp/197 bp	100	tetM	M21136
tet(O)	Y07780	188 bp/189 bp	99	tet(O)	Y07780
tet(S)	L09756	172 bp/172 bp	100	tet	L09756
tet(X)	M37699	186 bp/186 bp	100	tetX	M37699
acrB	M94248	159 bp/160 bp	99	acrB	M94248
acrD	U12598	185 bp/185 bp	100	acrD	U12598
mexB	L11616	142 bp/147 bp	97	ttgB	CT573326
		130 bp/147 bp	88	mexB	L11616
mexD	U57969	182 bp/182 bp	100	mexD	U57969
mexD	NC_003430	184 bp/185 bp	99	mexD	NC_003430
mexF	X99514	348 bp/348 bp	100	mexF	AE004091
		345 bp/348 bp	99	mexF	X99514
mexI	AE004837	170 bp/170 bp	100	mexI	AE004837 (new accession no. AE004091)
mexY	AB015853	198 bp/198 bp	100	mexY	AB015853

*In some cases the two best hits are given: (i) best hit to a related gene; (ii) hit to the reference gene.

from bacteria of the WWTP's final effluents confirms that these determinants are released into the environment, which might facilitate further dissemination among environmental bacteria. Moreover, it appeared that wastewater purification processes operating within the WWTP analysed are not appropriate to significantly reduce the spectrum of resistance genes that are detectable in the final effluents.

The composition of the plasmid pool analysed was biased, since plasmids were isolated from bacteria showing reduced susceptibility to different antibiotics. Accordingly, future projects will aim at the detection of antibiotic-resistance determinants in whole-community plasmid DNA preparations. In this context the microarray technology seems to be very well suited for simultaneous detection of hundreds of resistance determinants in samples derived from different WWTPs. Likewise, it would be informative to compare plasmid samples obtained from WWTPs that receive effluents from hospitals with those that are not connected to any medical facilities.

REFERENCES

1. Akinbowale, O. L., Peng, H. & Barton, M. D. (2007). Diversity of tetracycline resistance genes in bacteria from aquaculture sources in Australia. J Appl Microbiol 103, 2016–2025.
2. Altschul, S. F., Gish, W., Miller, W., Myers, E. W. & Lipman, D. J. (1990). Basic local alignment search tool. J Mol Biol 215, 403–410. CrossRefMedline
3. Bennett, P. M. (1999). Integrons and gene cassettes: a genetic construction kit for bacteria. J Antimicrob Chemother 43, 1–4.
4. Bönemann, G., Stiens, M., Pühler, A. & Schlüter, A. (2006). Mobilizable IncQ-related plasmid carrying a new quinolone resistance gene, qnrS2, isolated from the bacterial community of a wastewater treatment plant. Antimicrob Agents Chemother 50, 3075–3080.
5. Bonnet, R., Recule, C., Baraduc, R., Chanal, C., Sirot, D., De Champs, C. & Sirot, J. (2003). Effect of D240G substitution in a novel ESBL CTX-M-27. J Antimicrob Chemother 52, 29–35.
6. Bywater, R. J. (2004). Veterinary use of antimicrobials and emergence of resistance in zoonotic and sentinel bacteria in the EU. J Vet Med B Infect Dis Vet Public Health 51, 361–363.
7. Bywater, R. J. (2005). Identification and surveillance of antimicrobial resistance dissemination in animal production. Poult Sci 84, 644–648.
8. Cartelle, M., del Mar Tomas, M., Molina, F., Moure, R., Villanueva, R. & Bou, G. (2004). High-level resistance to ceftazidime conferred by a novel enzyme, CTX-M-32, derived from CTX-M-1 through a single Asp240-Gly substitution. Antimicrob Agents Chemother 48, 2308–2313.
9. Chee-Sanford, J. C., Aminov, R. I., Krapac, I. J., Garrigues-Jeanjean, N. & Mackie, R. I. (2001). Occurrence and diversity of tetracycline resistance genes in lagoons and

groundwater underlying two swine production facilities. Appl Environ Microbiol 67, 1494–1502.

10. Dang, H., Ren, J., Song, L., Sun, S. & An, L. (2008). Diverse tetracycline resistant bacteria and resistance genes from coastal waters of Jiaozhou Bay. Microb Ecol 55, 237–246.

11. Davies, J. (1994). Inactivation of antibiotics and the dissemination of resistance genes. Science 264, 375–382.

12. Davison, J. (1999). Genetic exchange between bacteria in the environment. Plasmid 42, 73–91.

13. Doublet, B., Weill, F. X., Fabre, L., Chaslus-Dancla, E. & Cloeckaert, A. (2004). Variant Salmonella genomic island 1 antibiotic resistance gene cluster containing a novel 3'-N-aminoglycoside acetyltransferase gene cassette, aac(3)-Id, in Salmonella enterica serovar newport. Antimicrob Agents Chemother 48, 3806–3812.

14. Eichenlaub, R., Figurski, D. & Helinski, D. R. (1977). Bidirection replication from a unique origin in a mini-F plasmid. Proc Natl Acad Sci U S A 74, 1138–1141.

15. Fiett, J., Baraniak, A., Mrowka, A., Fleischer, M., Drulis-Kawa, Z., Naumiuk, L., Samet, A., Hryniewicz, W. & Gniadkowski, M. (2006). Molecular epidemiology of acquired-metallo-beta-lactamase-producing bacteria in Poland. Antimicrob Agents Chemother 50, 880–886.

16. Geisenberger, O., Ammendola, A., Christensen, B. B., Molin, S., Schleifer, K. H. & Eberl, L. (1999). Monitoring the conjugal transfer of plasmid RP4 in activated sludge and in situ identification of the transconjugants. FEMS Microbiol Lett 174, 9–17.

17. Göbel, A., Thomsen, A., McArdell, C. S., Alder, A. C., Giger, W., Theiss, N., Löffler, D. & Ternes, T. A. (2005). Extraction and determination of sulfonamides, macro-lides, and trimethoprim in sewage sludge. J Chromatogr A 1085, 179–189.

18. Golet, E. M., Strehler, A., Alder, A. C. & Giger, W. (2002). Determination of flu-oroquinolone antibacterial agents in sewage sludge and sludge-treated soil using accelerated solvent extraction followed by solid-phase extraction. Anal Chem 74, 5455–5462.

19. Golet, E. M., Xifra, I., Siegrist, H., Alder, A. C. & Giger, W. (2003). Environmental exposure assessment of fluoroquinolone antibacterial agents from sewage to soil. Environ Sci Technol 37, 3243–3249.

20. Goossens, H. (2005). European status of resistance in nosocomial infections. Che-motherapy 51, 177–181.

21. Gordon, D., Abajian, C. & Green, P. (1998). Consed: a graphical tool for sequence finishing. Genome Res 8, 195–202.

22. Gordon, D., Desmarais, C. & Green, P. (2001). Automated finishing with autofinish. Genome Res 11, 614–625.

23. Götz, A., Pukall, R., Smit, E., Tietze, E., Prager, R., Tschäpe, H., van Elsas, J. D. & Smalla, K. (1996). Detection and characterization of broad-host-range plasmids in environmental bacteria by PCR. Appl Environ Microbiol 62, 2621–2628.

24. Guillaume, G., Verbrugge, D., Chasseur-Libotte, M., Moens, W. & Collard, J. (2000). PCR typing of tetracycline resistance determinants (Tet A–E) in Salmonella enterica serotype Hadar and in the microbial community of activated sludges from

hospital and urban wastewater treatment facilities in Belgium. FEMS Microbiol Ecol 32, 77–85.

25. Hall, R. M. & Collis, C. M. (1995). Mobile gene cassettes and integrons – capture and spread of genes by site-specific recombination. Mol Microbiol 15, 593–600.

26. Hata, M., Suzuki, M., Matsumoto, M., Takahashi, M., Sato, K., Ibe, S. & Sakae, K. (2005). Cloning of a novel gene for quinolone resistance from a transferable plasmid in Shigella flexneri 2b. Antimicrob Agents Chemother 49, 801–803.

27. Henriques, I., Moura, A., Alves, A., Saavedra, M. J. & Correia, A. (2006a). Analysing diversity among β-lactamase encoding genes in aquatic environments. FEMS Microbiol Ecol 56, 418–429.

28. Henriques, I. S., Fonseca, F., Alves, A., Saavedra, M. J. & Correia, A. (2006b). Occurrence and diversity of integrons and β-lactamase genes among ampicillin-resistant isolates from estuarine waters. Res Microbiol 157, 938–947.

29. Heritier, C., Poirel, L. & Nordmann, P. (2004). Genetic and biochemical characterization of a chromosome-encoded carbapenem-hydrolyzing ambler class D beta-lactamase from Shewanella algae. Antimicrob Agents Chemother 48, 1670–1675.

30. Heuer, H., Szczepanowski, R., Schneiker, S., Pühler, A., Top, E. M. & Schlüter, A. (2004). The complete sequences of plasmids pB2 and pB3 provide evidence for a recent ancestor of the IncP-1β group without any accessory genes. Microbiology 150, 3591–3599.

31. Jacoby, G. A., Walsh, K. E., Mills, D. M., Walker, V. J., Oh, H., Robicsek, A. & Hooper, D. C. (2006). qnrB, another plasmid-mediated gene for quinolone resistance. Antimicrob Agents Chemother 50, 1178–1182.

32. Jarnheimer, P. A., Ottoson, J., Lindberg, R., Stenstrom, T. A., Johansson, M., Tysklind, M., Winner, M. M. & Olsen, B. (2004). Fluoroquinolone antibiotics in a hospital sewage line; occurrence, distribution and impact on bacterial resistance. Scand J Infect Dis 36, 752–755.

33. Kümmerer, K. (2003). Significance of antibiotics in the environment. J Antimicrob Chemother 52, 5–7.

34. Kümmerer, K., Al-Ahmad, A. & Mersch-Sundermann, V. (2000). Biodegradability of some antibiotics, elimination of the genotoxicity and affection of wastewater bacteria in a simple test. Chemosphere 40, 701–710.

35. Larkin, M. A., Blackshields, G., Brown, N. P., Chenna, R., McGettigan, P. A., McWilliam, H., Valentin, F., Wallace, I. M., Wilm, A. & other authors (2007). CLUSTAL W and CLUSTAL_X version 2.0. Bioinformatics 23, 2947–2948.

36. Lee, H. B., Peart, T. E. & Svoboda, M. L. (2007). Determination of ofloxacin, norfloxacin, and ciprofloxacin in sewage by selective solid-phase extraction, liquid chromatography with fluorescence detection, and liquid chromatography-tandem mass spectrometry. J Chromatogr A 1139, 45–52.

37. Lim, S. M. & Webb, S. A. (2005). Nosocomial bacterial infections in intensive care units. I. Organisms and mechanisms of antibiotic resistance. Anaesthesia 60, 887–902.

38. Lindberg, R. H., Wennberg, P., Johansson, M. I., Tysklind, M. & Andersson, B. A. (2005). Screening of human antibiotic substances and determination of weekly mass flows in five sewage treatment plants in Sweden. Environ Sci Technol 39, 3421–3429.

39. Lindberg, R. H., Olofsson, U., Rendahl, P., Johansson, M. I., Tysklind, M. & Andersson, B. A. (2006). Behavior of fluoroquinolones and trimethoprim during mechanical, chemical, and active sludge treatment of sewage water and digestion of sludge. Environ Sci Technol 40, 1042–1048.

40. Llanes, C., Gabant, P., Couturier, M., Bayer, L. & Plesiat, P. (1996). Molecular analysis of the replication elements of the broad-host-range RepA/C replicon. Plasmid 36, 26–35.

41. Mach, P. A. & Grimes, D. J. (1982). R-plasmid transfer in a wastewater treatment plant. Appl Environ Microbiol 44, 1395–1403.

42. Mancini, P., Fertels, S., Nave, D. & Gealt, M. A. (1987). Mobilization of plasmid pHSV106 from Escherichia coli HB101 in a laboratory-scale waste treatment facility. Appl Environ Microbiol 53, 665–671.

43. Marcinek, H., Wirth, R., Muscholl-Silberhorn, A. & Gauer, M. (1998). Enterococcus faecalis gene transfer under natural conditions in municipal sewage water treatment plants. Appl Environ Microbiol 64, 626–632.

44. Mazel, D. & Davies, J. (1999). Antibiotic resistance in microbes. Cell Mol Life Sci 56, 742–754.

45. Nakata, H., Kannan, K., Jones, P. D. & Giesy, J. P. (2005). Determination of fluoroquinolone antibiotics in wastewater effluents by liquid chromatography-mass spectrometry and fluorescence detection. Chemosphere 58, 759–766.

46. Nüßlein, K., Maris, D., Timmis, K. & Dwyer, D. F. (1992). Expression and transfer of engineered catabolic pathways harbored by Pseudomonas spp. introduced into activated sludge microcosms. Appl Environ Microbiol 58, 3380–3386.

47. Poirel, L., Marqué, S., Héritier, C., Segonds, C., Chabanon, G. & Nordmann, P. (2005). OXA-58, a novel class D β-lactamase involved in resistance to carbapenems in Acinetobacter baumannii. Antimicrob Agents Chemother 49, 202–208.

48. Prasher, D. C., Eckenrode, V. K., Ward, W. W., Prendergast, F. G. & Cormier, M. J. (1992). Primary structure of the Aequorea victoria green-fluorescent protein. Gene 111, 229–233.

49. Rowe-Magnus, D. A. & Mazel, D. (1999). Resistance gene capture. Curr Opin Microbiol 2, 483–488.

50. Rozen, S. & Skaletsky, H. (2000). Primer3 on the WWW for general users and for biologist programmers. Methods Mol Biol 132, 365–386.

51. Sambrook, J., Fritsch, E. F. & Maniatis, T. (1989). Molecular Cloning: a Laboratory Manual. Cold Spring Harbor, NY: Cold Spring Harbor Laboratory.

52. Schlüter, A., Heuer, H., Szczepanowski, R., Forney, L. J., Thomas, C. M., Pühler, A. & Top, E. M. (2003). The 64508 bp IncP-1β antibiotic multiresistance plasmid pB10 isolated from a waste-water treatment plant provides evidence for recombination between members of different branches of the IncP-1β group. Microbiology 149, 3139–3153.

53. Schlüter, A., Heuer, H., Szczepanowski, R., Poler, S. M., Schneiker, S., Pühler, A. & Top, E. M. (2005). Plasmid pB8 is closely related to the prototype IncP-1β plasmid R751 but transfers poorly to Escherichia coli and carries a new transposon encoding a small multidrug resistance efflux protein. Plasmid 54, 135–148.

54. Schlüter, A., Szczepanowski, R., Kurz, N., Schneiker, S., Krahn, I. & Pühler, A. (2007a). Erythromycin resistance-conferring plasmid pRSB105, isolated from a

218   Wastewater and Public Health: Bacterial and Pharmaceutical Exposures

sewage treatment plant, harbors a new macrolide resistance determinant, an integron-containing Tn402-like element, and a large region of unknown function. Appl Environ Microbiol 73, 1952–1960.

55. Schlüter, A., Krahn, I., Kollin, F., Bönemann, G., Stiens, M., Szczepanowski, R., Schneiker, S. & Pühler, A. (2007b). IncP-1β plasmid pGNB1 isolated from a bacterial community from a wastewater treatment plant mediates decolorization of triphenylmethane dyes. Appl Environ Microbiol 73, 6345–6350.

56. Schlüter, A., Krause, L., Szczepanowski, R., Goesmann, A. & Pühler, A. (2008). Genetic diversity and composition of a plasmid metagenome from a wastewater treatment plant. J Biotechnol 136, 65–76.

57. Schwartz, T., Kohnen, W., Jansen, B. & Obst, O. (2003). Detection of antibiotic-resistant bacteria and their resistance genes in wastewater, surface water, and drinking water biofilms. FEMS Microbiol Ecol 43, 325–335.

58. Seveno, N. A., Kallifidas, D., Smalla, K., van Elsas, J. D., Collard, J. M., Karagouni, A. D. & Wellington, E. M. H. (2002). Occurrence and reservoirs of antibiotic resistance genes in the environment. Rev Med Microbiol 13, 15–27.

59. Smith, M. S., Yang, R. K., Knapp, C. W., Niu, Y., Peak, N., Hanfelt, M. M., Galland, J. C. & Graham, D. W. (2004). Quantification of tetracycline resistance genes in feedlot lagoons by real-time PCR. Appl Environ Microbiol 70, 7372–7377.

60. Stiens, M., Becker, A., Bekel, T., Gödde, V., Goesmann, A., Niehaus, K., Schneiker-Bekel, S., Selbitschka, W., Weidner, S. & other authors (2008). Comparative genomic hybridisation and ultrafast pyrosequencing revealed remarkable differences between the Sinorhizobium meliloti genomes of the model strain Rm1021 and the field isolate SM11. J Biotechnol 136, 31–37.

61. Szczepanowski, R., Krahn, I., Linke, B., Goesmann, A., Pühler, A. & Schlüter, A. (2004). Antibiotic multiresistance plasmid pRSB101 isolated from a wastewater treatment plant is related to plasmids residing in phytopathogenic bacteria and carries eight different resistance determinants including a multidrug transport system. Microbiology 150, 3613–3630.

62. Szczepanowski, R., Braun, S., Riedel, V., Schneiker, S., Krahn, I., Pühler, A. & Schlüter, A. (2005). The 120592 bp IncF plasmid pRSB107 isolated from a sewage-treatment plant encodes nine different antibiotic-resistance determinants, two iron-acquisition systems and other putative virulence-associated functions. Microbiology 151, 1095–1111.

63. Szczepanowski, R., Krahn, I., Bohn, N., Pühler, A. & Schlüter, A. (2007). Novel macrolide resistance module carried by the IncP-1β resistance plasmid pRSB111, isolated from a wastewater treatment plant. Antimicrob Agents Chemother 51, 673–678.

64. Szczepanowski, R., Bekel, T., Goesmann, A., Krause, L., Krömeke, H., Kaiser, O., Eichler, W., Pühler, A. & Schlüter, A. (2008). Insight into the plasmid metagenome of wastewater treatment plant bacteria showing reduced susceptibility to antimicrobial drugs analysed by the 454-pyrosequencing technology. J Biotechnol 136, 54–64.

65. Tauch, A., Schlüter, A., Bischoff, N., Goesmann, A., Meyer, F. & Pühler, A. (2003). The 79,370 bp conjugative plasmid pB4 consists of an IncP-1β backbone loaded with a chromate resistance transposon, the strA-strB streptomycin resistance gene

pair, the oxacillinase gene blaNPS-1, and a tripartite antibiotic efflux system of the resistance-nodulation-division family. Mol Genet Genomics 268, 570–584.

66. Tennstedt, T., Szczepanowski, R., Braun, S., Pühler, A. & Schlüter, A. (2003). Occurrence of integron-associated resistance gene cassettes located on antibiotic resistance plasmids isolated from a wastewater treatment plant. FEMS Microbiol Ecol 45, 239–252.

67. Tennstedt, T., Szczepanowski, R., Krahn, I., Pühler, A. & Schlüter, A. (2005). Sequence of the 68,869 bp IncP-1α plasmid pTB11 from a waste-water treatment plant reveals a highly conserved backbone, a Tn402-like integron and other transposable elements. Plasmid 53, 218–238.

68. Toleman, M. A., Biedenbach, D., Bennett, D., Jones, R. N. & Walsh, T. R. (2003). Genetic characterization of a novel metallo-beta-lactamase gene, blaIMP-13, harboured by a novel Tn5051-type transposon disseminating carbapenemase genes in Europe: report from the SENTRY worldwide antimicrobial surveillance programme. J Antimicrob Chemother 52, 583–590.

69. Volkmann, H., Schwartz, T., Bischoff, P., Kirchen, S. & Obst, U. (2004). Detection of clinically relevant antibiotic-resistance genes in municipal wastewater using real-time PCR (TaqMan). J Microbiol Methods 56, 277–286.

70. Vourli, S., Giakkoupi, P., Miriagou, V., Tzelepi, E., Vatopoulos, A. C. & Tzouvelekis, L. S. (2004). Novel GES/IBC extended-spectrum β-lactamase variants with carbapenemase activity in clinical enterobacteria. FEMS Microbiol Lett 234, 209–213.

71. Wassenaar, T. M. (2005). Use of antimicrobial agents in veterinary medicine and implications for human health. Crit Rev Microbiol 31, 155–169.

72. Xiong, J., Hynes, M. F., Ye, H., Chen, H., Yang, Y., M'Zali, F. & Hawkey, P. M. (2006). blaIMP-9 and its association with large plasmids carried by Pseudomonas aeruginosa isolates from the People's Republic of China. Antimicrob Agents Chemother 50, 355–358.

PART III

PHARMACEUTICAL
AND HORMONE EXPOSURE

CHAPTER 10

Simultaneous Determination of 24 Antidepressant Drugs and Their Metabolites in Wastewater by Ultra-High Performance Liquid Chromatography–Tandem Mass Spectrometry

LING-HUI SHENG, HONG-RUI CHEN, YING-BIN HUO, JING WANG, YU ZHANG, MIN YANG, AND HONG-XUN ZHANG

10.1 INTRODUCTION

In recent years, more and more human-use pharmaceuticals and their metabolites have been found in waters from municipal wastewater treatment plants (WWTPs) [1,2,3,4,5]. The antidepressant drugs are a class of neuroactive compounds that act as selective serotonin reuptake inhibitors and serotonin noradrenergic reuptake inhibitors, and some of their metabolites also retain the pharmacologic activity and are capable of contributing to serotonin reuptake inhibition [6,7,8]. The unaltered drugs or their main metabolites have been detected in municipal wastewaters in many countries [9,10,11,12,13,14,15,16].

Generally, the antidepressants exist in wastewater in low concentrations. However, their potent psychoactivity cannot be neglected. There are many reports on the potential effects of such compounds that may lead to physiological and behavioral changes in aquatic organisms and accumulate in their tissues [17,18,19,20,21]. Therefore, it is important to understand the environmental profile of these antidepressants and their metabolites in order to assess their potential ecological impact [22,23]. GÒmez et al. analyzed the effluents from three WWTPs in Spain using solid phase extraction (SPE) with an Oasis HLB sorbent, followed by gas chromatography-tandem mass spectrometry and found that the wastewater contained 8 ng/L of carbamazepine [24]. Lajeunesse et al. studied the basic antidepressants and their N-desmethyl metabolites in raw sewage and wastewater from Montreal (Canada). Six basic antidepressants and four of their metabolites were detected by SPE and LC–MS/MS in raw sewage and roughly primary-treated wastewater with concentrations ranging from 2–346 ng/L [14]. Chen et al. found primidone and carbamazepine in wastewater effluent samples in Hebei Province of China with maximum concentrations of 74 ng/L and 103 ng/L, respectively [25].

To our knowledge, only a few top-selling antidepressants found in wastewater were studied in different countries [9,10,11,12,13,14,15,16]. Simultaneous analysis of the most representative antidepressant drugs such as imipramine, nortriptyline, clozapine, mirtazapine, and their metabolites has not been reported up to now. The aim of this study is to investigate the occurrence of representative antidepressant drugs in wastewater from WWTPs in Beijing, China, which may facilitate a better understanding of the potential ecological and human health risks of these antidepressants and their metabolites in wastewater.

10.2 RESULTS AND DISCUSSION

10.2.1 OPTIMIZATION OF LC–MS/MS

It has been shown that mass spectrometry is the most suitable tool for determination of trace environmental pollutants due to its high selectivity and sensitivity. Because most antidepressant drugs contain nitrogen-containing moieties with high proton affinities, in this study positive electrospray ionization (+ESI) was employed to analyze these antidepressant pharmaceuticals. Full-scan and MS/MS mass spectra of each compound were obtained by a 5500 Qtrap Tandem Mass Spectrometer via infusion of 50 ng/L individual standard solutions at a flow rate of 10 iL/min. Two MRM transitions (quantifier and qualifier) were selected for each compound except norfluoxetine, fluoxetine and duloxetine, for which only one transition was monitored due to the poor fragmentation. The declustering potential (DP), collision energy (CE) and collision cell exit potential (CXP) for each analyte were optimized through direct infusion into mass spectrometer at the flow rate of 10 iL/min. The optimized parameters of each mass transition were listed in Table 1.

Since antidepressants and their metabolites exhibit basic characteristics due to their amine moieties, a source of protons in solution is essential to convert the basic substances into their cationic forms. In the present study, to improve ionization during the electrospray process, different concentrations of the mobile solutes ammonium formate and ammonium acetate in acetonitrile were tested. Meanwhile, the pH of the buffer was also evaluated. The highest signals sensitivity was observed when using ammonium acetate (10 mM, pH 3.0) as the mobile phase.

In order to get a good separation of all analytes, the retention and separation ability of CSH C18, HSS T3 C18 and BEH C18 columns were tested under the selected chromatographic conditions. The optimal peak shapes and resolution were achieved with BEH C18 column. The representative ion chromatograms of each compound obtained from a mixture of 30 standard samples are shown in Figure 1.

10.2.2 SOLID-PHASE EXTRACTION (SPE) STUDY

To extract the target analytes from aqueous samples and reduce coextractive impurities from matrixes, two kinds of SPE sorbents, i.e., Oasis HLB and 3 M Empore cation, were compared. The recovery percentage was used to estimate the SPE process efficiency for the selected analytes. In the case of HLB, which was used in most studies of trace drugs, the recovery was low, especially for olanzapine (44%) and clozapine (40%).

TABLE 1. Mass spectrometer parameters for each analyte.

Analyte	Precusorion (m/z)	Production (m/z)	Declustering potential (V)	Collision energy (V)	Collision Exit potential(V)
Citalopram	325.2	262.1 *	110	26	17
		234		36	
Sertraline	306.2	275 *	30	16	13.8
		159		34	
Venlafaxine	278.2	58 *	60	20	9
		121		34	
Desmethyl- venlafaxine	264.2	58 *	65	50.7	9
		107		49.7	
Fluvoxamine	319.2	71 *	30	18.3	16
		200		26.9	
Desmethyl Fluvoxamine	305.1	71 *	44	15	13
		200		32	
Fluoxetine	310	44 *	40	54	18.9
Norfluoxetine	296.2	134 *	37	8.5	11.2
Paroxetine	330.2	192 *	100	28	18
		70.2		50	8
Duloxetine	298.2	154 *	40	7.4	17
Mirtazapine	266.1	195 *	60	28.5	15.6
		72		26.2	
Desmeth- ylmirtazapine	252	195 *	60	30.9	17.9
		209		28.6	
Trazodone	372.3	148 *	80	42	14
		176		31	
Clomipramine	315.2	86 *	66	22.9	10
		58		70	
Norclomip- ramine	301.1	72 *	56	20	15
		242		30	
Clozapine	327.1	270 *	34	35	32.1
		296		35	
Demethylclo- zapine	313.1	270 *	100	33	15
		227		38	
Quetiapine	384.2	253 *	110	30	19
		221		50	

TABLE 1. CONTINUED.

Olanzapine	313.1	256.1 *	80	29.7	20
		198		50	
Imipramine	281.1	86 *	66	23.5	9
		58		59.8	
Bupropion	240.1	184 *	50	16.5	14
		131		35	
Amitriptyline	278.2	91 *	80	29	13
		233.2		23	
Milnacipran	274.1	100 *	30	25	12
		230		15	
Nortriptyline	264.2	91 *	100	27	10
		105		26	
Doxepin	280.2	107.1 *	58	28	13
		141		31.5	
Nordoxepin	266.1	107 *	83.4	24.9	13
		91		27	
Fluphenazine	438.2	171.1 *	60	32.3	12.2
		143.1		39.2	
Moclobemide	269.1	182 *	60	27.4	11
		139		39.6	
Chlorproma-zine	319.1	86 *	85	24	17
		58		53	
Mianserin	265.2	208 *	130	28	18
		118		35	

* Quantitative ion.

In contrast, the antidepressants and their metabolites were efficiently retained with higher recovery on 3 M cation disk. The higher recovery on 3 M cation disk was due to the filled cation-exchange sorbents which provide more effective sample clean-up for basic drugs by protonation of the free basic group. In the present study, the relative recovery rate for each

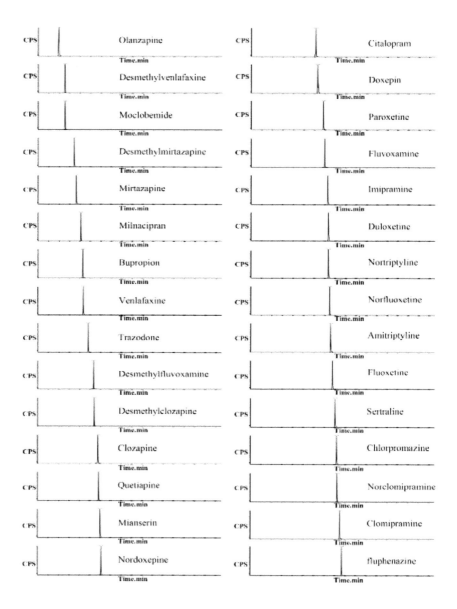

Figure 1. MRM chromatogram of 30 compounds obtained from a mixture of standard samples.

target compound was from 81.2% to 118% at 1 ng/mL. Therefore, the 3M cation disk was chosen for SPE because of its superior extraction recoveries for all analytes in the wastewater.

10.2.3 METHOD VALIDATION

The method was validated for linearity, limit of detection (LOD) and limit of quantification (LOQ) by the analysis of spiked milli-Q water samples. The results are listed in Table 2. In each case, a weighted linear regression line was applied. The LOD and LOQ were calculated as three and ten times signal to noise ratio at the 1 ng/mL concentration with an accuracy ranged from 80% to 120% and precision within $\pm$ 15% of the target concentration. The LODs ranged from 0.01 to 0.15 ng/mL and the LODs of norclomipramine, clozapine, quetiapine, nordoxepin and chlorpromazine were the lowest (0.01 ng/mL). The determination coefficient (r^2) of all calibration curves was more than 0.99 within in the tested concentration range (0.1–25 ng/mL). Both selectivity and sensitivity of the established method were satisfactory, and no interfering substances presented at the appropriate retention times.

10.2.4 APPLICATION TO ENVIRONMENTAL ANALYSIS

This method was applied to detect antidepressants and their metabolites in the wastewater from three WWTPs in Beijing. The compounds were detected and confirmed by comparing their retention time and MRM transitions Eight antidepressants and two metabolites were successfully found in all wastewater samples. The concentrations of the antidepressants and metabolites are listed in Table 3. The metabolite desmethylvenlafaxine was found at the highest concentration of (415.6 $\pm$ 32.9) ng/L, while the desmethylmirtazapine was detected in the effluents of WWTPs at the lowest concentration of (4.0 $\pm$ 0.2) ng/L. The measured concentrations of these antidepressants and their metabolites were consistent with the previous reports about wastewaters [14,26].

TABLE 2. Correlation coefficient, linearity range, LOD and LOQ of the method.

Analyte	Correlation coefficient	Concentration range (ng/mL)	LOD (ng/mL)	LOQ (ng/mL)
Citalopram	0.9988	0.1–10	0.04	0.11
Sertraline	0.9989	0.1–10	0.02	0.06
Venlafaxine	0.9987	0.1–10	0.03	0.09
Desmethylvenlafaxine	0.9979	0.1–10	0.05	0.15
Fluvoxamine	0.9992	0.1–10	0.07	0.23
Desmethylfluvoxamine	0.9998	0.5–25	0.09	0.32
Fluoxetine	0.9986	0.1–10	0.12	0.50
norfluoxetine	0.9998	0.5–25	0.15	0.51
Paroxetine	0.9990	0.1–10	0.05	0.16
Duloxetine	0.9980	0.1–10	0.04	0.12
Mirtazapine	0.9989	0.1–10	0.02	0.06
Desmethylmirtazapine	0.9989	0.1–10	0.03	0.10
Trazodone	0.9995	0.1–10	0.02	0.06
Clomipramine	0.9995	0.1–10	0.03	0.08
Norclomipramine	0.9985	0.1–10	0.01	0.04
Clozapine	0.9976	0.1–10	0.01	0.04
Demethylclozapine	0.9986	0.1–10	0.05	0.16
Quetiapine	0.9991	0.1–10	0.01	0.03
Olanzapine	0.9973	0.1–10	0.05	0.17
Imipramine	0.9996	0.1–10	0.02	0.07
Bupropion	0.9997	0.1–10	0.04	0.13
Amitriptyline	0.9971	0.1–10	0.03	0.10
Milnacipran	0.9983	0.1–10	0.07	0.21
Nortriptyline	0.9985	0.1–10	0.03	0.10
Doxepin	0.9949	0.1–10	0.04	0.13
Nordoxepin	0.9977	0.1–10	0.01	0.04
Fluphenazine	0.9975	0.1–10	0.03	0.10
Moclobemide	0.9993	0.1–10	0.05	0.15
Chlorpromazine	0.9962	0.1–10	0.01	0.03
Mianserin	0.9986	0.1–10	0.04	0.14

10.3 EXPERIMENTAL

10.3.1 CHEMICALS AND REAGENTS

All solvents were of HPLC grade obtained from Merck (Darmstadt, Germany). Ammonium acetate and ammonium formate were purchased from Sigma–Aldrich (St. Louis, MO, USA). Chemical standards of the antidepressant drugs (citalopram, sertraline, venlafaxine, fluvoxamine, fluoxetine, paroxetine, duloxetine, mirtazapine, trazodone, clomipramine, clozapine, quetiapine, olanzapine, imipramine, bupropion, amitriptyline, doxepin, fluphenazine, moclobemide, mianserin, milnacipran) were purchased from National Institutes for Food and Drug Control of China (Beijing, China). Desmethyl -venlafaxine was purchased from Zibo Dingjin Chemical Co., Ltd (Shandong, China). Norfluoxetine, desmethylmirtazapine, desmethylvenlafaxine, desmethylfluvoxamine, chlorpromazine, demethylclozapine, norclomipramine, nortriptyline were purchased from Toronto Research Chemicals Inc. (North York, ON, Canada). Standard stock solutions of these compounds were prepared in methanol (1 mg/mL). Deionized water was purified using a Milli-Q water purification system (Merck Millipore, Bedford, MA, USA).

Water samples were collected between December 2012 and April 2013 from wastewater treat plants of Beijing. The water samples were vacuum filtered through 1 μm glass fiber filters, followed by 0.22 μm nylon membrane filters right after the sampling, and stored on −40 °C until the analysis.

10.3.2 APPARATUS AND OPERATION CONDITIONS

The chromatographic separations were performed using the Agilent 1290 UHPLC system (Agilent, Waldbronn, Germany) equipped with a BEH C18 column, 2.1 × 150 mm, 1.7 μm (Waters Corporation, Milford, MA, USA). Gradient elution was applied using 10 mM ammonium acetate containing 0.1% formic acid (A) and acetonitrile (B) as the mobile phase and programmed as follows: the gradient started with 20% eluent B in 5 min

TABLE 3. Correlation coefficient, linearity range, LOD and LOQ of the method.

Analyte	WWTP1#	WWTP2#	WWTP3#
Citalopram	20.6 ± 2.0	18.3 ± 2.0	28.8 ± 3.5
Sertraline	37.3 ± 1.3	40.6 ± 4.0	18.8 ± 2.8
Venlafaxine	31.8 ± 4.3	63.7 ± 2.6	30.3 ± 4.6
Desmethylvenlafaxine	52.3 ± 2.7	71.3 ± 4.8	415.6 ± 32.9[a]
Mirtazapine	62.8 ± 4.8	76.3 ± 5.8	84.2 ± 13.7
Desmethylmirtazapine	4.0 ± 0.2	7.3 ± 1.3	10.1 ± 1.5
Clomipramine	101.7 ± 11.6 [a]	77.5 ± 4.3	92.3 ± 10.7
Clozapine	91.7 ± 4.3	53.4 ± 4.8	163.9 ± 17.7[a]
Imipramine	10.8 ± 0.6	10.6 ± 2.1	10.9 ± 1.1
Nortriptyline	47.8 ± 1.3	35.1 ± 5.1	43.8 ± 7.8

a: The concentration was calculated from a diluted (1:10) sample.

and increased linearly up to 60% in 15 min. Then, it decreased linearly again down to 20% in 5 min. The flow rate was 0.3 mL/min.

All the samples were analyzed with a 5500Qtrap Tandem Mass Spectrometer (Applied Bioscience, Foster City, CA, USA). Quantification was achieved by using multiple reaction monitoring (MRM). The declustering potential (DP), collision energy (CE) and collision cell exit potential (CXP) for each compound were optimized, which are listed in Table 1. Additional instrumental parameters for all analytes were as follows: curtain gas (CUR) setting at 45 psi, spray voltage at 4,500 V, source temperature at 600 °C, gas1 at 60 psi and gas2 at 50 psi.

10.3.3 SAMPLE EXTRACTION

To remove suspended material, the water samples were vacuum filtered through 1 μm glass fiber filters, followed by 0.22 μm nylon membrane filters right after the sampling. The pH of each 300 mL of filtered sewage was adjusted to around 3 with hydrochloric acid (6 mol/L). An automated solid-phase extraction system Dex4790 (Horizon Technology, Salem, NH, USA) and 47 mm cation disks (3M Corporation, St. Paul, MN, USA) were employed. SPE was performed at flow rates 30–40 mL/min. Before the sample loading, the disk was washed with 8 mL of methanol, 8 mL of Milli-Q water, and 8 mL of acidified (pH 3) water. Samples were loaded onto the disk by gravity, and then the disk was washed with 10 mL of Milli-Q water, 10 mL 10% methanol and vacuum dried for 5 min. Finally the target drugs were eluted using 12 mL of 8% ammonia solution in methanol. The eluate was evaporated to dryness at 40 °C under a gentle stream of nitrogen and reconstituted in 1 mL of 20% acetronitrile solution.

10.4 CONCLUSIONS

A method based on the application of UHPLC–MS/MS for the simultaneous determination of 24 antidepressants and six metabolites in WWTP wastewaters was developed and validated. This is the first time the determination of up to 30 psychoactive compounds in one injection is reported, and 10 compounds were determined in all the effluents from three WWTPs in Beijing. Of these analytes, four antidepressants and two metabolites (venlafaxine, desmethylvenlafaxine, mirtazapine, desmethylmirtazapine, imipramine, nortriptylin) were found for the first time. High sensitivity and lower LOD were obtained thanks to the use of characteristic transitions. This study revealed the prevalence of antidepressants and their metabolites in wastewater effluents. The method developed in this study proved to be a valuable tool in the analytical characterization of antidepressants and their metabolites in wastewater, and may be helpful for determination of these drugs in sediment samples.

REFERENCES

1.   Baker, D.R.; Kasprzyk-Hordern, B. Spatial and temporal occurrence of pharmaceuticals and illicit drugs in the aqueous environment and during wastewater treatment: New developments. Sci. Total Environ. 2013, 454, 442–456.

2. Kovalova, L.; Siegrist, H.; Singer, H.; Wittmer, A.; McArdell, C.S. Hospital wastewater treatment by membrane bioreactor: Performance and efficiency for organic micropollutant elimination. Environ. Sci. Technol. 2012, 46, 1536–1545.

3. Jelic, A.; Fatone, F.; di Fabio, S.; Petrovic, M.; Cecchi, F.; Barcelo, D. Tracing pharmaceuticals in a municipal plant for integrated wastewater and organic solid waste treatment. Sci. Total Environ. 2012, 433, 352–361.

4. Jelic, A.; Gros, M.; Ginebreda, A.; Cespedes-Sánchez, R.; Ventura, F.; Petrovic, M.; Barcelo, D. Occurrence, partition and removal of pharmaceuticals in sewage water and sludge during wastewater treatment. Water Res. 2011, 45, 1165–1176.

5. Bisceglia, K.J.; Yu, J.T.; Coelhan, M.; Bouwer, E.J.; Roberts, A.L. Trace determination of pharmaceuticals and other wastewater-derived micropollutants by solid phase extraction and gas chromatography/mass spectrometry. J. Chromatogr. A 2010, 1217, 558–564.

6. DeVane, C.L. Metabolism and pharmacokinetics of selective serotonin reuptake inhibitors. Cell. Mol. Neurobiol. 1999, 19, 443–466. [Google Scholar] [CrossRef]

7. Anderson, H.D.; Pace, W.D.; Libby, A.M.; West, D.R.; Valuck, R.J. Rates of 5 common antidepressant side effects among new adult and adolescent cases of depression: A retrospective US claims study. Clin. Ther. 2012, 34, 113–123.

8. Henry, A.; Kisicki, M.D.; Varley, C. Efficacy and safety of antidepressant drug treatment in children and adolescents. Mol. Psychiat. 2011, 17, 1186–1193.

9. Mohapatra, D.P.; Brar, S.K.; Tyagi, R.D.; Picard, P.; Surampalli, R.Y. Carbamazepine in municipal wastewater and wastewater sludge: Ultrafast quantification by laser diode thermal desorption-atmospheric pressure chemical ionization coupled with tandem mass spectrometry. Talanta 2012, 99, 247–255.

10. Lajeunesse, A.; Smyth, S.A.; Barclay, K.; Sauvéc, S.; Gagnon, C. Distribution of antidepressant residues in wastewater and biosolids following different treatment processes by municipal wastewater treatment plants in Canada. Water Res. 2012, 46, 5600–5612.

11. Baker, D.R.; Očenášková, V.; Kvicalova, M.; Kasprzyk-Hordern, B. Drugs of abuse in wastewater and suspended particulate matter—Further developments in sewage epidemiology. Environ. Int. 2012, 48, 28–38.

12. Tarcomnicu, I.; van Nuijs, A.L.; Simons, W.; Bervoets, L.; Blust, R.; Jorens, P.G.; Neelsa, H.; Covaci, A. Simultaneous determination of 15 top-prescribed pharmaceuticals and their metabolites in influent wastewater by reversed-phase liquid chromatography coupled to tandem mass spectrometry. Talanta 2011, 83, 795–803.

13. Bahlmann, A.; Weller, M.G.; Panne, U.; Schneider, R.J. Monitoring carbamazepine in surface and wastewaters by an immunoassay based on a monoclonal antibody. Anal. Bioanal. Chem. 2009, 395, 1809–1820.

14. Lajeunesse, A.; Gagnon, C.; Sauvé, S. Determination of basic antidepressants and their N-Desmethyl metabolites in raw sewage and wastewater using solid-phase extraction and liquid chromatography–tandem mass spectrometry. Anal. Chem. 2008, 80, 5325–5333.

15. Yu, K.; Li, B.; Zhang, T. Direct rapid analysis of multiple PPCPs in municipal wastewater using ultrahigh performance liquid chromatography–tandem mass spectrometry without SPE pre-concentration. Anal. Chim. Acta 2012, 738, 59–68.

16. Aranas, A.T.; Guidote, A.M., Jr.; Haddad, P.R.; Quirino, J.P. Sweeping–micellar electrokinetic chromatography for the simultaneous analysis of tricyclic antidepressant and â-blocker drugs in wastewater. Talanta 2011, 85, 86–90.

17. Fong, P.P.; Molnar, N. Antidepressants cause foot detachment from substrate in five species of marine snail. Mar. Environ. Res. 2013, 84, 24–30.

18. Smith, E.M.; Iftikar, F.I.; Higgins, S.; Irshad, A.; Jandoc, R.; Lee, M.; Wilson, J.Y. In vitro inhibition of cytochrome P450-mediated reactions by gemfibrozil, erythromycin, ciprofloxacin and fluoxetine in fish liver microsomes. Aquat. Toxicol. 2012, 109, 259–266.

19. Berg, C.; Backström, T.; Winberg, S.; Lindberg, R.; Brandt, I. Developmental exposure to fluoxetine modulates the serotonin system in hypothalamus. PLoS One 2013, 8, e55053.

20. Backhaus, T.; Porsbring, T.; Arrhenius, Å.; Brosche, S.; Johansson, P.; Blanck, H. Single-substance and mixture toxicity of five pharmaceuticals and personal care products to marine periphyton communities. Environ. Toxicol. Chem. 2011, 30, 2030–2040.

21. Neuwoehner, J.; Escher, B.I. The pH-dependent toxicity of basic pharmaceuticals in the green algae Scenedesmus vacuolatus can be explained with a toxicokinetic ion-trapping model. Aquat. Toxicol. 2011, 101, 266–275.

22. Li, H.; Sumarah, M.W.; Topp, E. Persistence of the tricyclic antidepressant drugs amitriptyline and nortriptyline in agriculture soils. Environ. Toxicol. Chem. 2013, 32, 509–516.

23. Niemi, L.M.; Stencel, K.A.; Murphy, M.J.; Schultz, M.M. Quantitative determination of antidepressants and their select degradates by liquid chromatography/electrospray ionization tandem mass spectrometry in biosolids destined for land application. Anal. Chem. 2013, 85, 7279–7286.

24. Gómez, M.J.; Agüera, A.; Mezcua, M.; Hurtado, J.; Mocholí, F.; Fernández-Alba, A.R. Simultaneous analysis of neutral and acidic pharmaceuticals as well as related compounds by gas chromatography–tandem mass spectrometry in wastewater. Talanta 2007, 73, 314–320.

25. Chen, F.; Ying, G.G.; Kong, L.X.; Wang, L.; Zhao, J.L.; Zhou, L.J.; Zhang, L.J. Distribution and accumulation of endocrine-disrupting chemicals and pharmaceuticals in wastewater irrigated soils in Hebei, China. Environ. Pollut. 2011, 159, 1490–1498.

26. Metcalfe, C.D.; Chu, S.; Judt, C.; Li, H.; Oakes, K.D.; Servos, M.R.; Andrews, D.M. Antidepressants and their metabolites in municipal wastewater, and downstream exposure in an urban watershed. Environ. Toxicol. Chem. 2010, 29, 79–89.

CHAPTER 11

Removal and Transformation of Pharmaceuticals in Wastewater Treatment Plants and Constructed Wetlands

E. LEE, S. LEE, J. PARK, Y. KIM, AND J. CHO

11.1 INTRODUCTION

For several decades, pharmaceuticals and personal care products (PPCPs) have been noticed as emerging problematic compounds (Ternes et al., 1998; Snyder et al., 2003). In order to reduce residual concentrations of PPCPs, various advance treatments have been studied by many research groups (Lee and Gunten, 2010; Rosal et al., 2010 and many others). However, high levels of PPCPs are still detected in wastewater effluent, surface waters, and drinking waters (Kim et al., 2007; Lee et al., 2012; Yoon et al., 2010; Benotti et al., 2009). Since those trace organic compounds have been detected even in treated drinking waters by Benotti et al. (2009), control of micropollutants has been important especially in the wastewater treatment plant, the main source of the micropollutants in the aquatic environments.

Removal and Transformation of Pharmaceuticals in Wastewater Treatment Plants and Constructed Wetlands. © Lee, E., Lee, S., Park, J., Kim, Y., and Cho, J. Drink. Water Eng. Sci., 6, 89-98, doi:10.5194/dwes-6-89-2013, 2013. Creative Commons Attribution License 3.0.

Meanwhile, constructed wetlands have been introduced as an alternative to wastewater treatment for micropollutants removal (Matamoros and Bayona, 2006). And few reports have been focused on the relationship between log D and removal of PPCPs in wetland systems. In previous study (Lee et al., 2011), removal efficiency in constructed wetlands has been investigated using corresponding octanol-water partitioning coefficient of pharmaceuticals.

And few studies have been reported with regarding to behaviors of pharmaceutical metabolites in various environments (Stumpf et al., 1998; Quintana et al., 2005). The potential ecotoxicological effect of pharmaceutical metabolites is still not known, thus, there is necessity to study further about the fate of pharmaceutical metabolites in various wastewater treatment processes.

In this study, removal and transformation of pharmaceuticals has been investigated with regarding to physicochemical and structural properties of pharmaceuticals and their metabolites in various environments such as WWTPs and constructed wetlands receiving wastewater effluent.

11.2 MATERIALS AND METHODS

11.2.1 TARGET COMPOUNDS

Various micropollutant compounds, including 9 pharmaceuticals, 11 selected metabolites, and 1 personal care product, were selected in this study. Acetaminophen (ACT), atenolol (ATN), carbamazepine (CBZ), diclofenac (DCF), glimepiride (GMP), ibuprofen (IBU), naproxen (NPX), O-desmethyl-naproxen (O-desmethyl-NPX), sulfamethoxazole (SMX), and tri(2-chloroethyl) phosphate (TCEP) were obtained from Sigma-Aldrich (St. Louis, MO). Caffeine (CAF) was purchased via Fluka Chemie GmbH (Buchs, Switzerland). 1-hydroxy ibuprofen (IBU-1OH), 2-hydroxy ibuprofen (IBU-2OH), ibuprofen carboxylic acid (IBU-CA), iopromide (IOP), and N-acetyl-sulfamethoxazole (N-acetyl-SMX), paraxanthine, paraxanthine-1-methyl-d3, 4-hydroxy diclofenac were purchased via Tronto Research Inc (Tronto, Canada). Authentic carbamazepine metabolites 10, 11-dihydro-10, 11-epoxycarbamazepine (CBZ-EP),

2-hydroxycarbamazepine (CBZ-2OH), 3-hydroxycarbamazepine (CBZ-3OH), and 10, 11-dihydro-10-hydroxycarbamazepine (CBZ-10OH) were provided from Norvatis Pharma AG (Basel, Switzerland). Carbamazepine-d_{10} and $^{13}C_1$-naproxen-d_3 were purchased from Cambridge Isotope Laboratories (Andover, MA, US). Atenolol-d_7 and diclofenac-d_4 were obtained from C/D/N Isotopes (Pointe-Claire, Canada). Ibuprofen-d_3, N^4-acetylsulfamethoxazole-d_4, and sulfamethoxazole-d_4 were obtained from Tronto Research Inc (Tronto, Canada). HPLC grade methanol was obtained from J.T. Baker (Philipsburg, NJ, US). Methyl tert-butyl ether (MTBE) and formic acid were obtained from Sigma-Aldrich (St. Louis, MO, US). Information of target compounds and their metabolites is listed in Table 1. The log K_{ow}, log D_{ow}, and pK_a for investigated compounds was calculated based on the molecular structures using ChemAxon Marvin Calculator Plugin. The log D_{ow} is a pH dependent log Kow value and in this study log D at pH 7 was used.

11.2.2 SAMPLE COLLECTION

In June 2010, water samples were collected from each process (influent and effluents of the various unit operations) in three different wastewater treatment plant. One is Gwangju primary municipal wastewater treatment plant, which is operated with the nitrogen and phosphorus removal treatment system (600 000 m^3 day^{-1}). Second sampling sites was Gwangju secondary municipal wastewater treatment plant, which had two different treatment trains (Conventional Activated Sludge and Modified Ludzack-Ettinger (MLE) process), having total capacity up to 120 000 m^3 day^{-1}. In those two different Gwangju municipal WWTPs, sodium hypochlorite was added as a disinfectant and final effluent samples were collected after disinfection process. Additionally, water samples from both Damyang wastewater treatment plant and Damyang constructed wetlands were collected and studied. There was no disinfection system in Damyang WWTP and final effluent was collected after secondary treatment. Damyang constructed wetlands connected to Damyang wastewater treatment plant is free surface flow constructed wetlands, which have two different ponds, containing *Aporus* ponds followed by *Typha* ponds. The hydraulic retention time of wetlands is approximately 6 h and flow rate is 1800 m^3 day^{-1}. Every sample was spiked with a biocide sodium azide and ascorbic acids to quench any residual oxidant in the field.

11.2.3 SPE AND LC-MS/MS ANALYSIS

After filtration using glass fiber membrane filter, all analytes were extracted by using AutoTrace automated solid phase extraction (SPE) system (Caliper Corporation, Hopkington, MA), as depicted by Vanderford and Snyder (2006). Briefly, the 6 mL, 500 mg hydrophilic-lipophilic balance (HLB) glass cartridges (Waters Corporation, Milford, MA) were preconditioned in the following order: 5 mL of MTBE, 5 mL of methanol, and 5 mL of deionized water. 500 mL of samples, spiked with the addition of standards for internal calibration, were loaded onto the cartridges at 15 mL min^{-1} in duplicate, after which the cartridges were rinsed with 5 mL of deionized pure water, and then dried with a steam of air for 50 min. The cartridges were eluted with 5 mL of methanol, followed by 5 mL of 1/9 (v/v) methanol/MTBE. The eluted solution was concentrated in a water bath at 40 °C with a gentle stream of air to a final volume of 500 μL, which was a concentration factor of 1000. The levels of pharmaceuticals and their metabolites were then measured using a Water 2695 Separations Module (Waters, Milford, MA) coupled with a Micromass Quattro Micro triple quadrupole tandem mass spectrometer (Micromass, Manchester, UK) in electrospray ionization mode (ESI). A 20 μL sample loop and 150 × 2.1 mm SunFire C18 column with a particle size of 3.5 μm (Waters, Milford, MA) was employed for analyte separation. A binary gradient, consisting of 0.1 % formic acid (eluent A) and 100 % acetonitrile (eluent B), was used at a flow rate of 0.2 mL min^{-1}. Selected PPCPs and their metabolites was analyzed using two different gradients. The gradient used for the PPCPs and most of the metabolites was: gradient with 15 % of B was held for 4 min, increased linearly to 80 % for 6 min, held for 3 min with 80 % of B and then linearly increased to 100 % for 7 min. The gradient used for the carbamazepine metabolites was: gradient with 10 % of B increased linearly to 40 % for 15 min, increased linearly to 90 % for 10 min, and held until 30 min (Kang et al., 2008).

TABLE 1. Tested parent compounds and metabolites.

Analytes	Uses	Structure	pK$_a$^{a}	log K$_{ow}$^{a}	log D^a at pH7
Acetaminophen	Analgesic		9.46	0.91	0.91
Glimepiride	Anticholesterol		4.32	3.12	2.18
tri(2-chloroethyl) phosphate (TCEP)	Flame retardant		N.E	2.11	2.11
Carbamazepine (CBZ)	Anticonvulsant		N.E	2.77	2.77
CBZ-EP	Carbamazepine metabolite		N.E.	1.97	1.97

TABLE 1. CONTINUED.

CBZ-2OH	Carbamazepine metabolite		9.3	2.66	2.66
CBZ-3OH	Carbamazepine metabolite		9.46	2.66	2.66
CBZ-10OH	Carbamazepine metabolite		14.1	1.73	1.73
Sulfamethoxazole	Antibiotic		6.16	0.79	0.14
N4-acetylsulfamethoxazole	Sulfamethoxazole metabolite		5.88	0.86	0.1
Diclofenac	Analgesic		4.00	4.26	0.96

TABLE 1. CONTINUED.

4-hydroxyl-diclofenac	Diclofenac metabolite		3.76, 8.61	3.96	0.89
Naproxen	Analgesic		4.19	2.99	0.25
O-desmethyl-naproxen	Naproxen metabolite		4.34, 9.78	3.9	0.23
Ibuprofen	Analgesic		4.85	3.84	1.71
1-hydroxyl-ibuprofen	Ibuprofen metabolite		4.90	2.85	0.77

TABLE 1. CONTINUED.

2-hydroxyl-ibuprofen	Ibuprofen metabolite		4.63	2.08	0.03
Ibuprofen carboxylic acid	Ibuprofen metabolite		3.97, 4.77	2.78	-2.36
Caffeine	Stimulant		N.E	-0.55	-0.55
Paraxanthine	Caffeine metabolite		10.76	0.24	0.24

[a] log P, log D, and pK_a value were calculated from the Software Calculator Plugins.
[b] N.E: nonexistent at pH range 1–14.

A 5 min equilibration step with gradient of 10 % B was used at the beginning of each run. Detail LC-MS/MS analysis condition and analytical parameters of target compounds are shown in Table 2.

11.2.4 MOLECULAR ORBITAL CALCULATIONS

Molecular orbital were calculated single determinant (Hartree-Fock) for optimization bearing the minimum energy obtained at the AM1 level. All semi-empirical calculations to obtain the point charge and electron density for pharmaceuticals and metabolites were performed in MO-G with a SCIGRESS package version 7.7 (Fujitsu Co. Ltd.) (Watanabe et al., 2003).

11.3 RESULTS AND DISCUSSION

11.3.1 REMOVAL OF PHARMACEUTICALS AND PERSONAL CARE PRODUCTS IN WWTPS AND WETLANDS

Removal of PPCPs in municipal wastewater treatment plants and constructed wetlands were investigated and summarized by comparing the concentrations of PPCPs in the influent and final effluent of each WWTP in Table 3. Unfortunately, the contribution of different transport mechanisms such as biodegradation, adsorption to sludge and sediments, and oxidation during disinfection was not considered here. Gwangju WWTP was found as a major source, releasing tons of micropollutants to the Yeongsan River water, and, the concentrations of PPCPs in Gwangju WWTPs ($\sim\mu gL^{-1}$) were generally much higher than in Damyang WWTP ($\sim$ ng L^{-1}), presumably resulting from the dense populations in Gwangju area.

Particularly, high concentrations of caffeine and acetaminophen in WWTP influents reflect frequent use and ingestion of those compounds in this urban area. Most of PPCPs exhibited high removal efficiency (> 90 %) during the WWTPs processes, except for carbamazepine removal in Gwangju secondary WWTP (removal efficiency at 74 %).

TABLE 2. Analytical parameters of selected compounds (MDL: method detection limit; RL: reporting limit).

Compound	Retention time (min)	Cone voltage (V)	Collision energy (eV)	Parent ion (m/z)	Daughter ion (m/z)	MDL (ng L^{-1})	RL (ng L^{-1})
ESI negative							
Diclofenac	15.53	15	10	294	249	1.3	3.9
Diclofenac-d$_4$	11.23	15	12	298	254		
4-OH-DCF	13.70	20	12	310	266	5.1	15.3
Ibuprofen	15.70	15	8	205	161	1.2	3.6
Ibuprofen-d$_3$	11.63	15	8	208	164		
IBU-1OH	12.87	15	7	221	177	5.8	17.5
IBU-2OH	12.23	17	7	221	177	0.9	2.7
IBU-CA	12.27	12	5	235	191	2.3	6.8
Naproxen	9.84	10	8	229	185	3.3	10.0
Naproxen-d$_3$	9.83	10	6	233	189		
O-desmethyl-NPX	8.10	15	12	215	171	1.2	3.5
N-acetyl-SMZ	7.71	30	13	294	198	0.9	2.7
TCEP	9.12	30	16	285	161	7.9	23.8
ESI positive							
Acetaminophen	3.98	28	17	152	110	0.5	1.5
Acetaminophen-d$_4$	3.98	28	17	156	114		
Glimepiride	11.34	28	13	491	352	1.5	4.6
Sulfamethoxazole	7.83	30	18	254	156	0.6	1.7
Sulfamethoxazole-d$_4$	7.80	25	15	258	160		

TABLE 2. CONTINUED.

Caffeine	6.11	35	20	195	138	1.5	4.6
Paraxanthine	2.72	30	22	181	124	9.3	27.9
Paraxanthine-d_3	2.72	30	22	184	124		
Carbamazepine	19.31	35	18	237	194	0.7	2.1
Carbamazepine-d_{10}	8.74	35	18	247	204		
CBZ-EP	15.64	28	24	253	180	1.2	3.6
CBZ-2OH	14.34	35	20	253	210	13.0	3.9
CBZ-3OH	15.78	35	20	253	210	1.5	4.5
CBZ-10OH	13.63	30	20	255	194	0.2	0.5

TABLE 3. Concentrations of selected PPCPs in WWTPs and wetlands.

PPCPs	Gwangju Primary WWTP		Gwangju Secondary WWTP		Damyang WWTP and Constructed wetlands				
	WWTP influent (ng L⁻¹)	WWTP effluent (ng L⁻¹)	WWTP influent (ng L⁻¹)	WWTP effluent (ng L⁻¹)	WWTP influent (ng L⁻¹)	WWTP effluent (ng L⁻¹)	Acorus wetland (ng L⁻¹)	Typha wetland (ng L⁻¹)	Wetland effluent (ng L⁻¹)
Caffeine	72 471.2	<4.6	45 457.0	1857.8	36 880.6	<4.6	<4.6	47.0	<4.6
Carbamazepine	2085.4	108.3	1668.8	362.8	844.5	417.2	387.1	161.4	268.3
Sulfamethoxazole	6048.6	88.8	8092.9	70.7	409.7	27.4	17.9	7.4	17.4
Acetaminophen	162 159.9	1705.8	227 705.4	180.5	194 586.8	189.7	354.8	549.5	349.6
Ibuprofen	2626.0	<3.6	4576.5	166.6	1645.3	17.9	43.2	52.9	47.2
Naproxen	1140.5	<10.0	1778.5	10.7	203.5	<10.0	11.4	<10.0	<10.0
Diclofenac	2673.5	111.7	2702.8	<3.8	150.5	19.8	20.15	<3.8	8.9
Glimepiride	18 895.3	197.4	40 338.6	86.6	36.8	12.3	11.3	<4.6	14.4
TCEP	6011.8	320.3	4774.7	<23.8	2608	340.6	277.85	186.35	268.4

After discharge of WWTP effluent into connected constructed wetlands, levels of some PPCPs (carbamazepine, sulfamethoxaolze, diclofenac, and TCEP) slightly decreased. This result indicates that operation of wastewater stabilization ponds or wetlands would be helpful to prevent release of micropollutants into surface water. Our research group has previously been studied the Damyang constructed wetlands with respect to the control of organic micropollutants, and, reported the efficiency of wetland treatments depending on the wetland characteristics and properties of micropollutants (Park et al., 2009). Many other studies have also been supported the necessity of additional wetland treatments for micropollutants control (Matamoros and Bayona, 2006; Conkle et al., 2008).

11.3.2 TRANSFORMATION OF PHARMACEUTICALS AND FORMATION OF METABOLITES IN WWTPS

Table 4 shows occurrence of parent pharmaceuticals and their metabolites in WWTPs and wetlands system. In the WWTPs influent, most metabolites were detected at high level of concentrations with range of $100 \sim 10\ 000$ ng L^{-1} compared to WWTP effluents, indicating dominant transformation pathway of metabolite resulting from human body rather than microbial transformation. However, after passing the WWTPs, concentrations of most metabolites were observed higher than that of parent compounds, indicating the structural stability of the metabolites during the WWTP process relative to their parent compounds. The stability of metabolites may be related to the difference of the log D value between parent pharmaceutical compounds and pharmaceutical metabolites. As pharmaceuticals transformed into their metabolites, their log D value at pH 7 slightly decreased (Table 1). Considering the dominant removal mechanism of micropollutants in WWTP process (i.e., sorption and biodegradation), decrease of log D may result in the reduction of sorption of pharmaceutical metabolites on the sludge surface. Consequently, pharmaceutical metabolites would be detected at higher levels than parent compounds in WWTP effluents.

TABLE 4. Concentrations of parent pharmaceuticals and their metabolites in WWTPs and wetlands.

Parent compounds and their metabolites	Gwangju Primary WWTP		Gwangju Secondary WWTP		Damyang WWTP and Constructed wetlands				
	WWTP influent (ng L⁻¹)	WWTP effluent (ng L⁻¹)	WWTP influent (ng L⁻¹)	WWTP effluent (ng L⁻¹)	WWTP influent (ng L⁻¹)	WWTP effluent (ng L⁻¹)	Acorus wetland (ng L⁻¹)	Typha wetland (ng L⁻¹)	Wetland effluent (ng L⁻¹)
Caffeine	72471.2	<4.6	45457.0	<4.6	36880.6	<4.6	<4.6	47.0	<4.6
Paraxanthine	8215.3	33.9	10442.1	<27.9	3222.3	<27.9	<27.9	<27.9	<27.9
Sulfamethoxazole	6048.6	88.8	8092.9	166.6	409.7	27.4	17.9	7.4	17.4
N-acetyl-SMZ	5224.8	64.5	6224.2	72.4	152.6	4.9	<2.7	<2.7	5.3
Naproxen	1140.5	<10.0	1778.5	10.7	203.5	<10.0	11.4	<10.0	<10.0
O-desmethyl-NPX	191.5	31.7	125.0	31.1	245.0	9.3	<3.5	<3.5	7.75
Diclofenac	2673.5	111.7	2702.8	86.6	150.5	19.8	20.15	<3.8	8.9
4-OH-DCF	530.6	582.1	396.0	504.6	212.0	40.2	21.15	<15.3	<15.3

In case of sulfamethoxazole metabolite, even though the concentrations of N-acetyl-sulfamethoxazole were low enough in wastewater effluents, by considering retransformation of acetylated metabolite back into its parent compound, N-acetyl-sulfamethoxazole should be assumed as a pharmaceutically active parent compound (Gö- bel et al., 2005). And based on this reason, monitoring of N-acetyl-sulfamethoxazole is important and necessary for the effective control of micropollutants in aquatic environment

During the engineered constructed wetland treatments, there was no specific change of pharmaceutical metabolites in exception with ibuprofen metabolites. As can be seen in Fig. 1, ibuprofen was differently transformed depending on the treatment process. In WWTPs influent, 2-hydroxyibuprofen, the well-known human metabolite of ibuprofen was dominant, whereas, after the activated sludge treatment, concentrations of 1-hydroxyibuprofen indicate the highest level in all WWTPs and wetlands. Significant formation of 1-hydroxyibuprofen in WWTPs might be possibly explained by preferential microbial metabolism in activated sludge treatment process. In the additional experimental results conducted in river waters, completely different composition of ibuprofen metabolites was observed. 2-hydroxy-ibuprofen was found to be dominant in both river waters, Yeongsan River and Seomjin River (data are not shown). This might infer that 2-hydroxyibuprofen is much more persistent and stable than 1-hydroxyibuprofen and even than parent compound, ibuprofen, as previously mentioned by Weigel et al. (2004).

In contrast, carbamazepine did not exhibit any change in transformation behavior during WWTPs and wetland treatments (Fig. 2). All the selected carbamazepine metabolites, 10,11-epoxy-carbamazepine, 2-hydroxycarbamazepine, 3-hydroxycarbamazepine and 10-hydroxy-carbamazepine was detected in influents, effluents of WWTPs and wetlands. Even though the levels of 10,11-epoxy-carbamazepine was low, the eco-toxicological effect of CBZ-EP on the environment is worthy of further examination due to its pharmaceutically active property like its parent compound, carbamazepine. The most dominant carbamazepine metabolite was 10-hydroxy-carbamazepine. The predominant formation of CBZ-10OH might be due to the different distribution of electron density of atoms in carbamazepine.

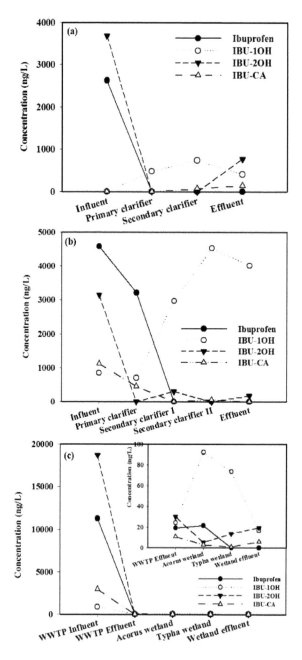

Figure 1. Transformation of ibuprofen in different WWTPs system: (a) Gwangju primary WWTP, (b) Gwangju secondary WWTP, (c) Damyang WWTP and constructed wetlands.

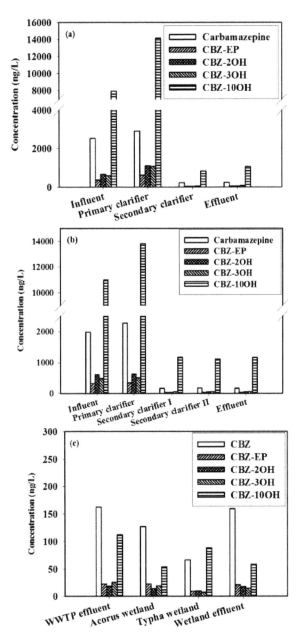

Figure 2. Distribution of carbamazepine metabolites in different WWTPs system: (a) Gwangju primary WWTP, (b) Gwangju secondary WWTP, (c) Damyang WWTP and constructed wetlands.

TABLE 5. Calculations of frontier electron density for carbamazepine.

Atom List	HOMO Destiny	Nucleophilic Frontier Destiny	Electrophilic Frontier Destiny	Radical Frontier Destiny	LUMO Destiny
C_1	0.148	0.202	0.215	0.208	0.131
C_2	0.119	0.252	0.175	0.214	0.169
C_3	0.064	0.222	0.108	0.165	0.136
C_4	0.118	0.113	0.184	0.148	0.057
C_5	0.046	0.201	0.088	0.145	0.090
C_6	0.067	0.113	0.136	0.124	0.049
N_7	0.074	0.003	0.211	0.107	0.000
C_8	0.039	0.141	0.080	0.111	0.054
C_9	0.010	0.117	0.036	0.076	0.030
C_{10}	0.068	0.214	0.111	0.163	0.133
C_{11}	0.012	0.088	0.044	0.066	0.013
C_{12}	0.046	0.094	0.117	0.105	0.047
C_{13}	0.024	0.048	0.076	0.062	0.007
C_{14}	0.111	0.129	0.173	0.151	0.072
C_{15}	0.016	0.043	0.078	0.060	0.010
C_{16}	0.012	0.010	0.021	0.016	0.000
N_{17}	0.007	0.002	0.072	0.037	0.000
O_{18}	0.018	0.002	0.064	0.033	0.000
H_{19}	0.000	0.001	0.001	0.001	0.000
H_{20}	0.000	0.000	0.001	0.001	0.000
H_{21}	0.001	0.002	0.002	0.002	0.001
H_{22}	0.002	0.001	0.003	0.002	0.000

11.3.3 EFFECT OF ELECTRON DENSITY ON TRANSFORMATION OF CARBAMAZEPINE INTO METABOLITES

In previous studies, electron density distribution of chemicals has been used to find the initial positions of OH radical attack in oxidation (Watanabe et

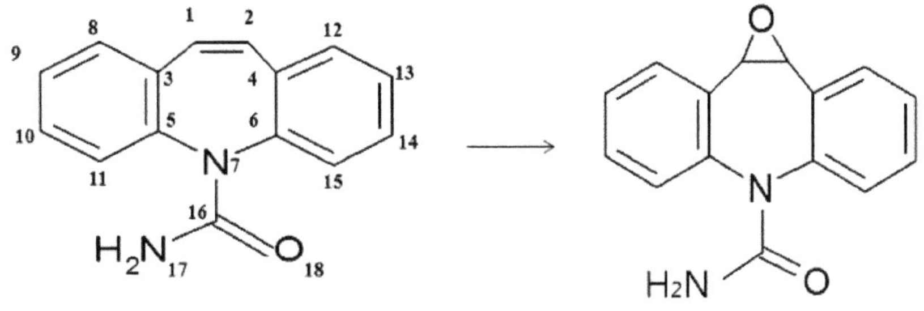

Figure 3. Structure of carbamazepine.

al., 2003; Jung et al., 2010; Heimstad et al., 2009). A higher electron density indicates more electrons in the bonds, resulting in electrophilic reaction (Horikoshi et al., 2004; Kaneco et al., 2006). In this study, electron density of pharmaceutical was examined to find the initial transformation position in various transformation mechanisms including biological process and photochemical oxidations. In previous study, Park et al. (2009) suggested that hydrolysis reaction at the amide and urea functional groups may lead to the biological transformation of carbamazepine. Frontier electron density of carbamazepine calculated by MOPAC, Scigress software is shown in Table 5. Radical frontier density, the averaged value of nucleophilic frontier density and electrophilic frontier density, are used in this study. Based on the electron density of carbamazepine, the carbon bonds between C_1 and C_2 showed the largest frontier electron density (0.208 and 0.214, respectively). This electron rich carbon bond of the olefin structure may provide the initial position of oxidation. Prevalent occurrence of specific carbamazepine metabolites such as CBZ-EP, CBZ- 10OH, and CBZ-DiOH in environment are presumably related with this oxidation pathway. However, other metabolites were difficult to find a relationship between transformation and electron density. The reason for this low relationship is thought that there might be some other different preferred transformation pathways depending on the intrinsic property of chemicals rather than electron density derived transformation. However, electron density of chemicals is still believed as a key parameter to elucidate unknown pathway in various processes, especially in oxidation process.

11.4 CONCLUSIONS

Removal and transformation of pharmaceuticals in WWTPs and constructed wetlands were extensively investigated in this study. Pharmaceuticals were effectively removed by different WWTP processes and wetlands. From this study, the additional operation of wastewater management wetlands was encouraged to prevent direct discharge of micropollutants into surface waters. And additionally, pharmaceutical metabolites were found to be more stable than the parent compounds during WWTP processes due to the lower log D value of metabolites than parent compounds. Different transformation pattern of pharmaceuticals was also observed, especially in transformation of ibuprofen. 1-hydroxyibuprofen was dominantly formed during biological treatment in WWTP, indicating preferential biotransformation of ibuprofen. At last, electron density of carbamazepine was examined to elucidate the transformation pathway. The electron rich $C_1=C_2$ bond in olefin structure of carbamazepine was revealed as an initial transformation position.

REFERENCES

1. Benotti, M. J., Trenholm, R. A., Vanderford, B. J., Holady, J. C., Stanford, B. D., and Snyder, S. A.: Pharmaceuticals and endocrine disrupting compounds in U.S. drinking water, Environ. Sci. Technol., 43, 597–603, 2009.
2. Conkle, J. L., White, J. R., and Metcalf, C. D.: Reduction of pharmaceutically active compounds by a lagoon wetland wastewater treatment system in Southeast Louisiana, Chemosphere, 73, 1741–1748, 2008.
3. Göbel, A., Thomsen, A., McArdell, C. A., Joss, A., and Giger, W.: Occurrence and sorption behavior of sulfonamides, macrolides, and trimethoprim in activated sludge treatment, Environ. Sci. Technol., 39, 3981–3989, 2005.
4. Heimstad, E. S., Bastos, P. M., Eriksson, J., Bergman, A., and Harju, M.: Quantitative structure – Photodegradation relationships of polybrominated diphenyl ethers, phenoxyphenols and selected organochlorines, Chemosphere, 77, 914–921, 2009.
5. Horikoshi, S., Tokunaga, A., Hidaka, H., and Serpone, N.: Environmental remediation by an integrated microwave/UV illumination method: VII. Thermal/non-thermal

effects in the microwaveassisted photocatalyzed mineralization of bisphenol-A, J. Photoch. Photobio. A, 162, 33–40, 2004.

6. Jung, Y. J., Oh, B. S., Kim, K. S., Koga, M., Shinohara, R., and Kang, J. W.: The degradation of diethyl phthalate (DEP) during ozonation: oxidation by-products study, J. Water Health., 8, 290– 298, 2010.

7. Kaneco, S., Katsumata, H., Suzuki, T., and Ohta, K.: Titanium dioxide mediated photocatalytic degradation of dibutyl phthalate in aqueous solution – kinetics, mineralization and reaction mechanism, Chem. Eng. J., 125, 59–66, 2006.

8. Kang, S.-I., Kang, S.-Y., and Hur, H.-G.: Identification of fungal metabolites of anticonvulsant drug carbamazepine, Appl. Microbiol. Biotechnol., 79, 663–669, 2008.

9. Kim, S. D., Cho, J., Kim, I. S., Vanderford, B. J., and Snyder, S. A.: Occurrence and removal of pharmaceuticals and endocrine disruptors in South Korean surface, drinking, and waste waters, Water Res., 41, 1013–1021, 2007.

10. Lee, E., Lee, S., Kim, Y., Huh, Y.-J., Kim, K.-S., Lim, B.-J., and Cho, J.: Wastewater Treatment Plant: Anthropogenic Micropollutant Indicators for Sustainable River Management, in: Encyclopedia of Sustainability Science and Technology, edited by: Meyers, R. A., Springer, 17, 11911–11932, 2012.

11. Lee, S., Kang, S. I., Lim, J. L., Huh, Y. J., and Cho, J.: Evaluating controllability of pharmaceuticals and metabolites, in biologically-engineered processes, using corresponding octanolwater partitioning coefficient with consideration of ionizable functional groups, Ecol. Eng., 37, 1595–1600, 2011.

12. Lee, Y. and von Gunten, U.: Oxidative transformation of micropollutants during municipal wastewater treatment: Comparison of kinetic aspects of selective (chlorine, chlorine dioxide, ferrateVI, and ozone) and non-selective oxidants (hydroxyl radical), Water Res., 44, 555–566, 2010.

13. Matamoros, V. and Bayona, J. M.: elimination of pharmaceuticals and personal care products in subsurface flow constructed wetlands, Environ. Sci. Technol., 40, 5811–5816, 2006.

14. Park, N., Vanderford, B. J., Snyder, S. A., Sarp, S., Kim, S. D., and Cho, J.: Effective controls of micropollutants included in wastewater effluent using constructed wetlands under anoxic condition, Ecol. Eng., 35, 418–423, 2009.

15. Quintana, J. B., Weiss, S., and Reemtsma, T.: Pathways and metabolites of microbial degradation of selected acidic pharmaceutical and their occurrence in municipal wastewater treated by a membrane bioreactor, Water Res., 39, 2654–2664, 2005.

16. Rosal, R., Rodríguez, A., Perdigón-Melón, J. A., Petre, A., GarcíaCalvo, E., Gómez, M. J., Agüera, A., and Fernández-Alba, A. R.: Occurrence of emerging pollutants in urban wastewaterand their removal through biological treatment followed by ozonation, Water Res., 44, 578–588, 2010.

17. Snyder, S. A., Westerhoff, P., Yoon, Y., and Sedlak, D. L.: pharmaceuticals, personal care products, and endocrine disruptors in water: implications for the water industry, Environ. Eng. Sci., 20, 449–469, 2003.

18. Stumpf, M., Ternes, T. A., Haberer, K., and Baumann, W.: Isolation of ibuprofenmetabolites and their importance as pollutants of the aquatic environment, Vom Wasser, 91, 291–303, 1998.

19. Ternes, T. A.: Occurrence of drugs in German sewage treatment plants and rivers, Water. Res., 32, 3245–3260, 1998.

20. Vanderford, B. J. and Snyder, S. A.: Analysis of pharmaceuticals in water by isotope dilution liquid chromatography/tandem mass spectrometry, Environ. Sci. Technol., 40, 7312–7320, 2006.

21. Watanabe, N., Horikoshi, S., Kawabe, H., Sugie, Y., Zhao, J., and Hidaka, H.: Photo-degradation mechanism for bisphenol A at the TiO2/H2O interfaces, Chemosphere, 52, 851–859, 2003.

22. Weigel, S., Berger, U., Jensen, E., Kallenborn, R., Thoresen, H., and Hühnerfuss, H.: Determination of selected pharmaceuticals and caffeine in sewage and seawater from Tromsø/Norway with emphasis on ibuprofen and its metabolites, Chemosphere, 56, 583– 592, 2004.

23. Yoon, Y., Ryu, J., Oh, J., Choi, B. G., and Snyder, S. A.: Occurrence of endocrine disrupting compounds, pharmaceuticals, and personal care products in the Han River (Seoul, South Korea), Sci. Total Environ., 408, 636–643, 2010.

CHAPTER 12

An Assessment of the Concentrations of Pharmaceutical Compounds in Wastewater Treatment Plants on the Island of Gran Canaria (Spain)

RAYCO GUEDES-ALONSO, CRISTINA AFONSO-OLIVARES, SARAH MONTESDEOCA-ESPONDA, ZORAIDA SOSA-FERRERA, AND JOSÉ JUAN SANTANA-RODRÍGUEZ

12.1 INTRODUCTION

Many modern pollution problems are a result of the intermittent or continuous release of chemical substances into the environment. Their presence is one of the main emerging issues that the organisations committed to public and environmental health have to address (Hernando et al. 2006a). Pharmaceutical compounds within this group of pollutants have raised increasing concerns over the last two decades because their effects on the environment are unknown. Thousands of tons of pharmaceuticals are used every year, in both human and veterinary medicine, and are released to the environment through metabolic excretion and improper disposal techniques. These compounds are not completely degraded at the wastewater treatment plants, and many of them are discharged into the environment through many sources and pathways (Wick et al. 2009).

These pharmaceutical compounds are objects of evaluation for their potential effects on aquatic organisms (Sanderson et al. 2004) and non-target species (Fent et al. 2006). The monitoring of these pharmaceuticals is therefore required to provide a greater knowledge with respect to their occurrence, their distribution in the environment and what effects they have on organisms when these organisms are exposed to low levels of pharmaceutical compounds (Pal et al. 2010).

The quantification of pharmaceuticals in human biological matrices such as blood, plasma or urine has been developed over a long period of time (Erny and Cifuentes 2006). Nevertheless, there is a greater difficulty in quantifying pharmaceuticals found in complex environmental samples because the concentrations of these compounds are very low and there are many compounds that can be quantified.

Gas chromatography (GC) and liquid chromatography (LC) are the most common techniques used to monitor the concentrations of organic contaminants in the environment (Hernando et al. 2006b; Zhang et al. 2011; Busetti et al. 2006; Gómez et al. 2007). Polar, non-volatile or thermally degradable compounds and their derivatives cannot be analysed by GC, and LC is an essential tool for the analysis of these types of compounds (Chen et al. 2008; Castiglioni et al. 2005). Liquid chromatography with tandem-mass spectrometry (LC-MS/MS) is the most commonly used technique (Wu et al. 2008; Baranowska and Kowalski 2010; Gros et al. 2012).

The low concentrations of pharmaceutical compounds in environmental samples render the employment of pre-treatment procedures such as the preconcentration and purification of these compounds to be necessary. The most common technique used to extract and preconcentrate pharmaceutical compounds present in environmental water samples is solid-phase extraction (SPE) (Pavlović et al. 2010; Afonso-Olivares et al. 2012; Montesdeoca-Esponda et al. 2012).

In this work, we present a monitoring of three groups of pharmaceutical compounds in wastewater samples. Group 1 consists of ketoprofen, naproxen, bezafibrate and carbamazepine, Group 2 consists of metamizole, atenolol, paraxanthine and fluoxetine and a third group consists of five fluoroquinolones, namely levofloxacin, norfloxacin, ciprofloxacin, enrofloxacin and sarafloxacin. Table 1 shows the structures and characteristics of the selected compounds. To do the monitoring, we have used the SPE, the LC-MS/MS and the UHPLC-MS/MS procedures that have previously been optimised by our group (Afonso-Olivares et al. 2012; Montesde-oca-Esponda et al. 2012). The selection of these pharmaceutical compounds was mainly based on the consumption of these compounds by the population.

The water effluent samples were collected bimonthly between January 2011 and December 2011 from two different wastewater treatment plants (WWTPs) located on Gran Canaria Island in Spain. The first WWTP (denoted as WWTP1) used conventional activated sludge method for the treatment of wastewater, while the other plant (WWTP2) employed a membrane bioreactor system for wastewater treatment.

12.2 MATERIALS AND METHODS

12.2.1 REAGENTS

All the pharmaceutical compounds and fluoroquinolones used were purchased from Sigma–Aldrich (Madrid, Spain). Stock solutions containing 1000 mg·L^{-1} of each analyte were prepared by dissolving the compound in methanol, and the solutions were stored in glass-stoppered bottles at 4°C prior to use. Working aqueous standard solutions were prepared daily. Ultrapure water was provided by a Milli-Q system (Millipore, Bedford, MA, USA). HPLC-grade methanol, LC-MS methanol, and LC-MS water as well as the formic acid and the ammonium formate used to adjust the pH of the LC-MS and UHPLC-MS mobile phases were obtained from Panreac Química (Barcelona, Spain). Polyoxyethylene 10 lauryl ether (POLE) was obtained from Sigma-Aldrich (Madrid, Spain) and prepared in Milli-Q water.

12.2.2 SAMPLE COLLECTION

Water samples were collected bimonthly from the effluent of two wastewater treatment plants located in the northern part of Gran Canaria in

TABLE 2. List of pharmaceutical compounds, identification number, pKa values, chemical structure and retention times.

Group of compounds	Identification number	Compound	pK$_a$	Structure	t$_R$ (min)
1	1	Naproxen	5.24		4.30
	2	Carbamaze-pine	13.9		2.14
	3	Ketoprofen	4.45		3.24
	4	Bezafibrate	-		5.52
2	5	Atenolol	9.64		7,23
	6	Metamizole	-		8,26
	7	Paraxanthine	8.5		11,23
	8	Fluoxetine	8.8		17,97

TABLE 2. CONTINUED.

3	9	Levofloxacin	-		2,03
	10	Norfloxacin	6.4		2,23
	11	Ciprofloxacin	5.9		2,61
	12	Enrofloxacin	-		3,16
	13	Sarafloxacin	-		4,76

2011. WWTP1 utilised a conventional activated sludge treatment system, while WWTP2 employed a membrane bioreactor treatment system. The samples were collected in 2 L amber glass bottles that were rinsed beforehand with methanol and water. Samples were purified through filtration with fibreglass filters and 0.65 μm membrane filters (Millipore, Ireland). The samples were stored in the dark at 4°C and extracted within 48 hours. Influent samples were not analysed, so the degradation of the compounds during treatment was not evaluated.

12.2.3 INSTRUMENTATION

The analysis of all pharmaceutical compounds except fluoroquinolones was performed in a Varian system (Varian Inc., Madrid, Spain), which consisted of a 320-MS LC/MS/MS system (triple quadrupole) equipped with an electrospray ionisation (ESI) interface, two pumps and a column valve module with an internal oven and an autosampler. The software used to control the system was MS Varian LC/MS Workstation Version 6.9.

The housing and desolvation temperatures were set at 60 and 250°C, respectively, for optimisation. Nitrogen was used as a nebuliser and a drying gas. Nebulisation was conducted at a pressure of 30 psi, and drying was conducted at a pressure of 65 psi. The capillary voltage was set to 4.5 kV in the positive mode (ESI+) and −3 kV in the negative mode (ESI−). The shield was programmed at −600/600 V (ESI+/ESI−), and the cone voltage was optimised for each compound. Collision-induced dissociation (CID) was conducted with argon as the collision gas at 1.94 psi.

The analysis of fluoroquinolones was performed in a UHPLC system from Waters (Madrid, Spain) consisting of an ACQUITY Quaternary Solvent Manager (QSM) used to load samples and wash and recondition the extraction column, an ACQUITY Binary Solvent Manager (BSM) for the elution of the analytes, a column manager, a 2777 autosampler equipped with a 25 μL syringe and a tray to hold 2 mL vials, and a ACQUITY tandem triple quadrupole (TQD) mass spectrometer with an electrospray ionization (ESI) interface. All Waters components (Madrid, Spain) were controlled using the MassLynx Mass Spectrometry Software. The electrospray ionisation parameters were fixed as follows: the capillary voltage was 3 kV, the cone voltage was 50 V, the source temperature was 120°C, the desolvation temperature was 450°C, and the desolvation gas flow rate was 800 L/hr. Nitrogen was used as the desolvation gas, and argon was employed as the collision gas.

The detailed MS/MS detection parameters for each pharmaceutical compound are presented in Table 2 and were optimised by the direct injection of a 1 mg·L^{-1} standard solution of each analyte into the detector at a flow rate of 10 μL·min^{-1}.

TABLE 2. Mass spectrometer parameters for the determination of target analytes.

N°	Compound	Precursor ion (m/z)	Capillary voltage (Ion mode)	Quantification ion, m/z (collision potential, V)	Quantification ion, m/z (collision potential, V)
1	Naproxen	231.2	36 (ESI +)	153.1 (28.5)	170.0 (22.0)
2	Carbamazepine	237.1	40 (ESI +)	194.0 (13.5)	192.0 (17.0)
3	Ketoprofen	255.1	52 (ESI +)	209.0 (10.0)	104.9 (18.5)
4	Bezafibrate	359.8	64 (ESI -)	273.7 (15.5)	153.5 (28.5)
5	Atenolol	267.0	52 (ESI +)	145.0 (23.5)	190.0 (16.5)
6	Metamizole	218.0	30 (ESI +)	56.0 (12.5)	97.0 (11.5)
7	Paraxanthine	181.0	40 (ESI +)	124.0 (17.0)	
8	Fluoxetine	310.0	30 (ESI +)	44.0 (6.5)	148.0 (5.5)
9	Levofloxacin	362.3	40 (ESI +)	318.3 (20.0)	261.2 (30.0)
10	Norfloxacin	320.3	40 (ESI +)	302.3 (20.0)	276.2 (15.0)
11	Ciprofloxacin	332.3	40 (ESI +)	314.3 (22.0)	288.2 (18.0)
12	Enrofloxacin	360.3	40 (ESI +)	316.3 (20.0)	245.3 (25.0)
13	Sarafloxacin	386.3	40 (ESI +)	368.3 (20.0)	299.2 (30.0)

12.2.4 CHROMATOGRAPHIC CONDITIONS

For Group 1 (ketoprofen, naproxen, bezafibrate and carbamazepine), the chromatographic column used was a 2.0 mm × 50 mm, Pursuit UPS C18 column with a particle size of 2.4 μm. The mobile phase used was a mixture of water containing 0,2% formic acid and 5 mM ammonium formate at a pH of 2.6 and methanol. A gradient programme started the elution at a 50:50 v/v mixture of water–methanol, which changed to 40:60 (v/v) for 9 minutes, following which it returned to the initial ratio in the next minute and stayed calibrating for another minute. The flow rate was 0.2 mL·min^{-1}, and the injection volume of the analyte was 10 μL.

For group 2 (metamizole, atenolol, paraxanthine and fluoxetine), the chromatographic column was a 3.0 mm × 100 mm, Sunfire™ C18 column with a particle size of 3.5 μm. The mobile phases, flow rate and injection volume used were the same as what was used with the analysis of group 1 compounds. The water:methanol gradient was started at 90:10 v/v. It changed to 45:55 v/v for 13 minutes and then to 35:65 in the next minute. Finally, the gradient was changed to 90:10 v/v after 16 minutes and stayed calibrating for another 5 minutes.

For the third group (levofloxacin, norfloxacin, ciprofloxacin, enrofloxacin and sarafloxacin), the analytical column was a 50 mm × 2.1 mm, AC-QUITY UHPLC BEH Waters C18 column with a particle size of 1.7 μm (Waters Chromatography, Barcelona, Spain) operating at a temperature of 40°C. The mobile phases were water, adjusted to a pH of 2.5 with 0.1% v/v formic acid, and methanol. The analysis was performed in isocratic mode, using a 50:50 v/v water–methanol mixture at a flow rate of 0.3 mL·min⁻¹. The sample volume injected was 10 μL.

12.2.5 SOLID-PHASE EXTRACTION

SPE conditions were optimised in previous studies (Afonso-Olivares et al. 2012; Montesdeoca-Esponda et al. 2012). Cartridges were conditioned with 5 ± 0.05 mL of methanol and 5 ± 0.05 mL of Milli-Q water at a flow-rate of 5 mL · min⁻¹ before each run. The sample was then passed through the cartridge at a flow of 10 mL · min⁻¹. A wash step was conducted using 5 ± 0.05 mL of Milli-Q water to remove any impurities. The cartridges were dried under vacuum for 10 minutes, and the analytes were eluted at an approximate flow rate of 1 mL · min⁻¹.

For Groups 1 and 2, the SPE cartridge used was an OASIS HLB 6 mL/200 mg cartridge (Waters, Spain). The sample volume was 250 ± 0.15 mL at a pH of 8.00 ± 0.01 and contained 0% w/v of sodium chloride. The desorption volume was 2 ± 0.02 mL of methanol. The eluents were then evaporated under a gentle nitrogen stream and reconstituted with 1 ± 0.01 mL of LC-MS grade water. These operating conditions for SPE allowed the samples to be preconcentrated by a factor of 250.

For Group 3, the SPE cartridge used was an OASIS HLB 6 mL/200 mg cartridge (Waters, Spain). The sample volume was 200 ± 0.15 mL at a pH of 3.00 ± 0.01 and contained 0% w/v of sodium chloride, and the desorption volume was 1 ± 0.01 mL of polyoxyethylene 10 lauryl ether (POLE) (Montesdeoca-Esponda et al. 2012). These operating conditions for SPE allowed the fluoroquinolones to be preconcentrated by a factor of 200.

12.3 RESULTS AND DISCUSSION

12.3.1 ANALYTICAL PARAMETERS

An external calibration was used for the quantification of the analytes by diluting the stock solution to six concentrations ranging between 1 and 500 $\mu g \cdot L^{-1}$, where each point corresponds to the mean value obtained from three area measurements. Analysis was conducted by LC-MS/MS for Group 1 and Group 2 compounds, and UHPLC-MS/MS was used for the fluoroquinolone group. Linear calibration plots for each analyte (r2 > 0.99) were obtained based on their chromatographic peak areas.

The limit of detection (LOD) and the limit of quantification (LOQ) for each compound were calculated from the signal to noise ratio of each individual peak in wastewater samples spiked with the analytes. The LOD was defined as the lowest concentration that gave a signal to noise ratio that was equal to 3. The LOQ was defined to be the lowest concentration that gave a signal to noise ratio that was equal to 10. The LODs ranged from $0.3 - 7.9$ ng $\cdot L^{-1}$ for Groups 1 and 2 and $5.3 - 11.1$ ng $\cdot L^{-1}$ for the fluoroquinolones. The LOQs for Group 1 and Group 2 ranged from $1.1 - 26.3$ ng $\cdot L^{-1}$, and they ranged from 17.7 to 37.0 ng $\cdot L^{-1}$ for the fluoroquinolones. Only Fluoxetine presented LOD and LOQ higher (97.4 and 324.7 respectively) because the transitions of fluoxetine presents more noise, so, the relation between signal and noise is lower, increasing the detection and quantification limits.

The performance and reliability of the process was studied by determining the repeatability of the quantification results for all target analytes under the described conditions. Six replicate samples were employed, obtaining relative standard deviations (RSDs) lower than 11% in all cases,

TABLE 3. Analytical parameters for the SPE-LC-MS/MS and SPE-UHPLC-MS/MS methods.

N°	Compound	RSD[a] (%) n=6	LOD[b] (ng/L)	LOQ[c] (ng/L)	Recovery (%) n=6
1	Naproxen	9.6	0.6	1.8	101.8 ± 7.0
2	Carbamazepine	10.7	0.3	1.1	105.6 ± 4.4
3	Ketoprofen	9.2	2.4	7.9	98.6 ± 9.4
4	Bezafibrate	7.8	2.9	9.6	91.6 ± 11.4
5	Atenolol	6.5	7.9	26.3	67.2 ± 4.4
6	Metamizole	7.9	6.3	21.1	54.4 ± 4.3
7	Paraxanthine	10.8	2.2	7.8	96.4 ± 10.4
8	Fluoxetine	7.7	97.4	324.7	21.0 ± 1.6
9	Levofloxacin	8.5	9.1	30.3	82.4 ± 14.0
10	Norfloxacin	8.5	8.5	28.0	85.3 ± 5.2
11	Ciprofloxacin	6.8	8.6	28.7	86.2 ± 2.1
12	Enrofloxacin	7.0	5.3	17.7	94.0 ± 6.1
13	Sarafloxacin	9.8	11.1	37.0	86.1 ± 11.2

[a]Relative Standard Derivation.
[b]Detection limits, calculated as signal to noise ratio of three times.
[c]Quantification limits, calculated as signal to noise ratio of ten times.

indicating a good repeatability. Finally, the recoveries of the SPE methods were measured in 6 real samples and they were over 67%, except for metamizole and fluoxetine (54 and 21% respectively). Table 3 shows the analytical parameters obtained for all compounds analysed.

12.3.2 ANALYSIS OF SELECTED COMPOUNDS IN WASTEWATER SAMPLES

The SPE extraction procedure was combined with the LC-MS/MS and the UHPLC-MS/MS detection methods for monitoring wastewater effluents from two different WWTPs located on Gran Canaria Island in Spain.

The samples were collected once every two months over the duration of a year. The first plant (WWTP1) uses the conventional activated sludge method for the treatment of wastewater, while the second plant (WWTP2) employs a membrane bioreactor (MBR) system for wastewater treatment. Both WWTPs operate at similar daily influent sewage volumetric flow rates (500 m3/day for WWTP1 and 700 m3/day for WWTP2) and treat the wastewater from similarly sized populations (5,000 inhabitants for WWTP1 and 7,000 inhabitants for WWTP2). Figure 1, demonstrate the MRM chromatograms corresponding to wastewater samples from WWTP1 that contain compounds from Groups 1 and 2 respectively. The results of the measurements are shown in Table 4.

We can observe that the concentrations of the group 1 compounds range consistently from 0.05 and 0.30 $\mu g \cdot L^{-1}$ for naproxen, carbamazepine and ketoprofen. Bezafibrate exhibits concentrations ranging between 0.04

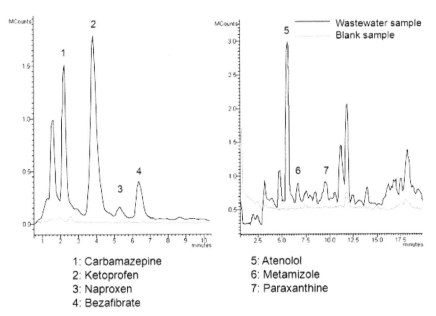

1: Carbamazepine
2: Ketoprofen
3: Naproxen
4: Bezafibrate

5: Atenolol
6: Metamizole
7: Paraxanthine

Figure 1. Chromatogram of WWTP1 sample with LC/MS-MS detection for Groups 1 and 2 of compounds.

and 2.15 $\mu g \cdot L^{-1}$. WWTP1 has higher naproxen, ketoprofen and bezafibrate concentrations that WWTP2 and a similar effluent concentration for carbamazepine.

There are more notable differences in the analysis of the pharmaceuticals in Group 2. In WWTP1, atenolol concentrations range between 0.04 and 0.95 $\mu g \cdot L^{-1}$, except in one sample (July 2011), where the concentration was 2.95 $\mu g \cdot L^{-1}$. Metamizole concentrations range between 0.25 and 3.45 $\mu g \cdot L^{-1}$, while the concentrations of paraxanthine are higher, ranging between 8.36 and 34.81 $\mu g \cdot L^{-1}$. The higher detected concentrations of paraxanthine can be explained through the fact that this compound is a metabolite of caffeine in the human body. In WWTP2, the concentrations of atenolol and metamizole are lower, except for metamizole in the May 2011 sample. In WWTP2, paraxanthine was not detected at all, while fluoxetine was not detected in either of the WWTPs.

The concentrations of the fluoroquinolones in both WWTPs are similar, ranging between 2.93 and 14.1 $\mu g \cdot L^{-1}$ for levofloxacin and between 11.1 and 20.3 $\mu g \cdot L^{-1}$ for ciprofloxacin. Norfloxacin, enrofloxacin and sarafloxacin were not detected in either of the WWTPs.

In summary, the concentrations of the pharmaceuticals and antibiotics detected at the wastewater treatment plant that operated with a membrane bioreactor treatment system are lower than that of the WWTP operating with a traditional technique such as activated sludge. Therefore, if the influent water quality of both the WWTPs is similar, the membrane bioreactor technique (MBR) can be said to be more efficient than the activated sludge technique.

12.4 CONCLUSIONS

A survey on the presence of pharmaceutical compounds in two wastewater treatment plants on the island of Gran Canaria in Spain was conducted. The scope of this study included eight common pharmaceutical compounds (naproxen, carbamazepine, ketoprofen, bezafibrate, atenolol, metamizole, paraxanthine and fluoxetine) and five fluoroquinolones (levofloxacin, norfloxacin, ciprofloxacin, enrofloxacin and sarafloxacin). Wastewater effluent samples were collected bimonthly in 2011. During

TABLE 4. Concentrations in µg·L⁻¹ found in treated water samples from two wastewater treatment plants of Gran Canaria island[a]

WWTP	Date	Naproxen	Carbamazepine	Ketoprofen	Bezafibrate
WWTP1	Jan-2011	0.06 ± 0.01	0.51 ± 0.01	0.78 ± 0.08	2.15 ± 0.37
	Mar-2011	0.10 ± 0.03	0.06 ± 0.01	0.22 ± 0.06	0.51 ± 0.05
	May-2011	0.08 ± 0.01	0.02 ± 0.00	0.26 ± 0.03	1.50 ± 0.02
	July-2011	0.25 ± 0.00	0.19 ± 0.01	1.36 ± 0.03	nd[b]
	Sept-2011	nd[b]	nd[b]	0.07 ± 0.01	0.04 ± 0.01
	Nov-2011	0.15 ±0.01	0.04 ± 0.00	0.14 ± 0.00	0.37 ± 0.07
WWTP2	Jan-2011	nd[b]	0.97 ± 0.03	0.11 ± 0.01	nd[b]
	Mar-2011	nd[b]	0.36 ± 0.03	nd[b]	nd[b]
	May-2011	0.05 ± 0.01	0.08 ± 0.00	0.30 ± 0.01	2.13 ± 0.08
	July-2011	nd[b]	0.67 ± 0.06	0.06 ± 0.00	nd[b]
	Sept-2011	nd[b]	0.19 ± 0.02	0.05 ± 0.01	nd[b]
	Nov-2011	0.21 ± 0.02	0.24 ± 0.00	0.05 ± 0.00	nd[b]

WWTP	Date	Atenolol	Metamizole	Paraxanthine	Fluoxetine
WWTP1	Jan-2011	0.31 ± 0.01	3.45 ± 2.33	12.31 ± 0.83	nd[b]
	Mar-2011	0.61 ± 0.20	0.41 ± 0.15	8.36 ± 0.02	nd[b]
	May-2011	0.65 ± 0.03	nd[b]	nd[b]	nd[b]
	July-2011	2.95 ± 0.03	0.80 ± 0.01	nd[b]	nd[b]
	Sept-2011	0.95 ± 0.12	1.65 ± 0.21	34.81 ± 1.64	nd[b]
	Nov-2011	0.04 ± 0,00	0.25 ± 0.08	nd	nd[b]

TABLE 4. CONTINUED.

WWTP	Date	Levofloxacin	Norfloxacin	Ciprofloxacin	Enrofloxacin	Sarafloxacin
WWTP2	Jan-2011	nd[b]	nd[b]	nd[b]	nd[b]	nd[b]
	Mar-2011	0.07 ± 0.01	1.19 ± 1.06	nd[b]	nd[b]	nd[b]
	May-2011	nd[b]	8.25 ± 0.19	nd[b]	nd[b]	nd[b]
	July-2011	0.12 ± 0.01	0.24 ± 0.02	nd[b]	nd[b]	nd[b]
	Sept-2011	0.26 ± 0.04	0.62 ± 0.14	nd[b]	nd[b]	nd[b]
	Nov-2011	0.04 ± 0.00	0.25 ± 0.08	nd[b]	nd[b]	nd[b]
WWTP1	Jan-2011	4.40 ± 0.20	nd[b]	nd[b]	nd[b]	nd[b]
	Mar-2011	2.93 ± 0.17	nd[b]	11.1 ± 0.75	nd[b]	nd[b]
	May-2011	3.70 ± 0.28	nd[b]	20.3 ± 1.81	nd[b]	nd[b]
	July-2011	nd[b]	nd[b]	nd[b]	nd[b]	nd[b]
	Sept-2011	nd[b]	nd[b]	nd[b]	nd[b]	nd[b]
	Nov-2011	0.44 ± 0.03	nd[b]	nd[b]	nd[b]	nd[b]
WWTP2	Jan-2011	5.90 ± 0,59	nd[b]	nd[b]	nd[b]	nd[b]
	Mar-2011	14.1 ± 0.92	nd[b]	nd[b]	nd[b]	nd[b]
	May-2011	6.24 ± 0.28	nd[b]	16.02 ± 0.82	nd[b]	nd[b]
	July-2011	nd[b]	nd[b]	nd[b]	nd[b]	nd[b]
	Sept-2011	nd[b]	nd[b]	nd[b]	nd[b]	nd[b]
	Nov-2011	nd[b]	nd[b]	nd[b]	nd[b]	nd[b]

[a]n = 3.
[b]nd = not detected.

the monitoring period, 9 analytes were detected in all samples, with an-algesics, anti-inflamatories and lipid regulators being the most frequently detected compounds.

A group of fluoroquinolones was selected for analysis because they were considered "priority pollutants" due to their potential hazardous effects on the aquatic environment.

The results show that the elimination of most of the analysed compounds is incomplete, but the membrane bioreactor technique is the more efficient of the two wastewater treatment process analysed in the removal of pharmaceutical compounds, and it results in lower effluent concentrations for most of the compounds in comparison with the activated sludge technique.

The results obtained in this monitoring work support the motivation for including pharmaceutical compounds in the monitoring of wastewater effluent quality.

REFERENCES

1. Afonso-Olivares C, Sosa-Ferrera Z, Santana-Rodríguez JJ (2012) Analysis of anti-inflammatory, analgesic, stimulant and antidepressant drugs in purified water from wastewater treatment plants using SPE-LC tandem mass spectrometry. J Environ Sci Heal A 47:887-895
2. Baranowska I, Kowalski B (2010) The development of SPE procedures and an UH-PLC method for the simultaneous determination of ten drugs in water samples. Water Air Soil Poll 211:417-425
3. Busetti F, Heitz A, Cuomo M, Badoer S, Traverso P (2006) Determination of six-teen polycyclic aromatic hydrocarbons in aqueous and solid samples from an Italian wastewater treatment plant. J Chromatogr A 1102:104-115
4. Castiglioni S, Bagnati R, Calamari D, Fanelli R, Zuccato E (2005) A multiresidue analytical method using solid-phase extraction and high-pressure liquid chromatography tandem mass spectrometry to measure pharmaceuticals of different therapeutic classes in urban wastewaters. J Chromatogr A 1092:206-215
5. Chen H-C, Wang P-L, Ding W-H (2008) Using liquid chromatography-ion trap mass spectrometry to determine pharmaceutical residues in Taiwanese rivers and waste-waters. Chemosphere 72:863-869 PubMed Abstract | Publisher Full Text OpenURL
6. Erny GL, Cifuentes A (2006) Liquid separation techniques coupled with mass spectrometry for chiral analysis of pharmaceutical compounds. J Pharmaceut Biomed 40:509-515
7. Fent K, Weston AA, Caminada D (2006) Ecotoxicology of human pharmaceuticals. Aquat Toxicol 76:122-159

8. Gómez MJ, Martínez Bueno MJ, Lacorte S, Fernández-Alba AR, Agüera A (2007) Pilot survey monitoring pharmaceuticals and related compounds in a sewage treatment plant located on the Mediterranean coast. Chemosphere 66:993-1002

9. Gros M, Rodríguez-Mozaz S, Barceló D (2012) Fast and comprehensive multi-resiude analysis of a broad range of human and veterinary pharmaceuticals and some of their metabolies in surface and treated waters by ultra-high performance liquid chromatography coupled to quadrupole-linear ion trap tandem mass sectrometry. J Chromatogr A 1248:104-121

10. Hernando MD, Mezcua M, Fernández-Alba AR, Barceló D (2006) Environmental risk assessment of pharmaceutical residues in wastewater effluents, surface waters and sediments. Talanta 69:334-342

11. Hernando MD, Heath E, Petrovic M, Barceló D (2006) Trace-level determination of pharmaceutical residues by LC-MS/MS in natural and treated waters. A pilot-survey study. Anal Bioanal Chem 385:985-991

12. Montesdeoca-Esponda S, Sosa-Ferrera Z, Santana-Rodríguez JJ (2012) Comparison of solid phase extraction using micellar desorption combined with LC-FD and LC-MS/MS in the determination of antibiotic fluoroquinolone residues in sewage samples. J Liq Chromatogr Relat Technol.

13. Pal A, Gin KYH, Lin AY-C, Reinhard M (2010) Impacts of emerging organic contaminants on freshwater resources: Review of recent occurrence, sources, fate and effects. Sci Total Environ 408:6062-6069

14. Pavlović DM, Babić S, Dolar D, Ašperger D, Košutić K, Horvat AJM, Kaštelan-Macan M (2010) Development and optimization of the SPE procedure for determination of pharmaceuticals in water samples by HPLC-diode array detection. J Sep Sci 33:258-267

15. Sanderson H, Johnson DJ, Reitsma T, Brain RA, Wilson CJ, Solomon KR (2004) Ranking and prioritization of environmental risks of pharmaceuticals in surface waters. Regul Toxicol Pharm 39:158-183

16. Wick A, Fink G, Joss A, Siegrist H, Ternes TA (2009) Fate of beta blockers and psycho-active drugs in conventional wastewater treatment. Water Res 43:1060-1074

17. Wu C, Spongberg AL, Witter JD (2008) Use of solid phase extraction and liquid chromatography-tandem mass spectrometry for simultaneous determination of various pharmaceuticals in surface water. Int J Environ Anal Chem 88:1033-1048

18. Zhang H-C, Yu X-J, Yang W-C, Peng J-F, Xu T, Yin D-Q (2011) MCX based solid phase extraction combined with liquid chromatography tandem mass spectrometry for the simultaneous determination of 31 endocrine-disrupting compounds in surface water of Shangai. J Chromatogr B 879:2998-3004

AUTHOR NOTES

CHAPTER 1

Acknowledgments

We are grateful to the Medical Research Council of South Africa as well as the Govan Mbeki Research and Development Center of the University of Fort Hare for the financial support.

CHAPTER 2

Acknowledgments

We wish to acknowledge the Water Research Commission (WRC) of South Africa for funding this research.

CHAPTER 3

Acknowledgments

This work was financially supported by the Department of Environment Health Engineering, School of Health, Isfahan University of Medical Sciences. The authors also greatly appreciate the staff at the north municipal wastewater treatment plant who contributed to this study. We also thank Mr. Akbar Hassanzadeh, Department of Biostatistics and Epidemiology for assistance in data analysis.

CHAPTER 4

Acknowledgments

This work was supported by the Polish Ministry of Science and Higher Education (project no. NN 304 202137 to EK and NN 305 461139 to AL) and by the Institute of Oceanology, Polish Academy of Sciences (statutory task no. IV.4.2 to EK). The authors would like to thank A. Moura for valuable comments and suggestions to improve the quality of the paper.

CHAPTER 5

Competing Interests
Dr. Balcázar serves as an academic editor for PLOS ONE. This does not alter the authors' adherence to all the PLOS ONE policies on sharing data and materials. The other authors have declared that no competing interests exist.

Acknowledgments
We thank G.A. Jacoby, G. Prats, M. Muniesa, C. Torres and S. Torriani for providing us with bacterial strains harboring ARGs. We are also grateful to the operators at the WWTP for their assistance, to A. Sànchez for his technical assistance, to M. Petrovic for her help with antibiotic quantifications and to C. Borrego, A.S. Trebitz and J.R. Pratt for their helpful discussions.

Author Contributions
Conceived and designed the experiments: EM JJ JLB. Performed the experiments: EM. Analyzed the data: EM JLB. Contributed reagents/materials/analysis tools: JLB. Wrote the manuscript: EM.

CHAPTER 7

Competing Interests
The authors have declared that no competing interests exist.

Acknowledgments
We would like to thank Mr. Jinbao Yin for his kind help on data analysis.

Author Contributions
Conceived and designed the experiments: ZW XXZ BL CL AL. Performed the experiments: ZW KH YM PS. Analyzed the data: ZW XXZ KH YM PS. Contributed reagents/materials/analysis tools: XXZ BL CL AL. Wrote the manuscript: ZW XXZ BL AL.

CHAPTER 8

Conflicts of Interests
The authors declare no conflict of interest.

Acknowledgments
This study was financially supported by National Natural Science Foundation of China (51008153 and 51278240) and Environmental Protection Research Foundation of Jiangsu Province (China) (2012044 and 2012045).

Author Contributions
Conceived and designed the experiments: Kailong Huang, Xu-Xiang Zhang and Hongqiang Ren; Performed the experiments: Kailong Huang and Junying Tang; Analyzed the data: Kailong Huang, Xu-Xiang Zhang and Hongqiang Ren; Contributed reagents/materials/analysis tools: Xu-Xiang Zhang, Ke Xu and Hongqiang Ren; Wrote the paper: Kailong Huang, Xu-Xiang Zhang and Hongqiang Ren.

CHAPTER 9

Acknowledgment
The authors thank the Bioinformatics Resource Facility (BRF) at the CeBiTec of Bielefeld University for support regarding bioinformatics issues. The work was supported by the Landesamt für Natur, Umwelt und Verbraucherschuts NRW (Germany).

CHAPTER 10

Conflicts of Interests
The authors declare no conflict of interest.

Acknowledgments
This study was supported by Major Science and Technology Program for Water Pollution Control and Treatment (2012ZX07313-001-07) and Special Fund for Quality Supervision Research in the Public Interest (200910107).

Author Contributions

Hong-Xun Zhang, Min Yang and Yu Zhang participated in designing the study. Ling-Hui Sheng, Hong-Rui Chen, Ying-Bin Huo and Jing Wang conducted the study and collected the data. Ling-Hui Sheng and Hong-Rui Chen analyzed the data and wrote the manuscript.

CHAPTER 11

Acknowledgements

This research was supported by the National Research Foundation of Korea (NRF) grant funded by the Korea government (MEST) (No. 2012047029).

CHAPTER 12

Conflicts of Interests

The authors declare that they have no competing interests.

Acknowledgments

This work was supported by funds provided by the ACIISI of the Autonomous Government of the Canary Islands (Spain) under research project number SolSubC200801000254. We thank the Instituto Tecnológico de Canarias S.A. (ITC) and the CANARAGUA S.A. for their participation in the project.

Author Contributions

RGA, CAO and SME have conceived the experiments and ZSF and JJSR have supervised and approved them. The experimental work has been done by RGA, CAO and SME. Each one developed the extraction, analysis and quantification of a group of compounds under study. RGA wrote the first draft of the Manuscript and CAO and SME contributed to the writing of it. Finally, ZSF and JJSR made critical revisions of the Manuscript and all the authors reviewed and approved the final version of the Manuscript.

INDEX